Praise for **HOW TO MAKE A ZOMBIE**

"Frank Swain's gripping book, *How to Make a Zombie*, reads like a nonfiction version of a Stephen King novel – you'll stay up all night reading it with goose bumps and the lights on. This story of life, death, and the grey areas in between, and the scientific (and pseudoscientific) attempts at resuscitation, resurrection, and immortality, is the best account I've read of what has to be humanity's biggest fear (death) and the lengths we have gone to in order to circumvent its finality. I read it straight through in one setting, and so will you."

MICHAEL SHERMER, author of *Why People Believe Weird Things* and *The Believing Brain* and publisher of *Skeptic* magazine

"From attempts to reanimate animals from death to mind-control experiments and brain-hacking parasites, this delightfully macabre book explores the reality of zombie mythology. Science punk Frank Swain has pulled off a masterful feat in this broad-ranging and fascinating book. Braiiins!"

DR LEWIS DARTNELL, research fellow at the University of Leicester and author of *Life in the Universe* and *My Tourist Guide to the Solar System... and Beyond*

"Swain serves up a ghoulish treat – the real-life zombies of science and nature! Packed full of bizarre research and jaw-dropping tales, his book succeeds in being simultaneously entertaining, informative, and slightly unnerving, since it turns out that the zombies are, quite likely, you and I."

ALEX BOESE, author of *Elephants on Acid* and *Electrified Sheep*

HOW TO MAKE A

ZOMBIE

THE REAL LIFE (AND DEATH) SCIENCE OF

REANIMATION AND MIND CONTROL

FRANK SWAIN

A Oneworld Book

First published by Oneworld Publications 2013

Illustration credits
Prologue/Epilogue: Zombie hand © pated/Patrick Ellis/iStockphoto
Chapter 1: "Haitian voodoo ritual" photo by Jerry Cooke
© Time & Life Pictures/Getty Images
Chapter 2: "Doctors Transferring Blood" © Bettmann/Corbis
Chapter 3: "Woman leaning over laboratory table" by George Skadding
© Time & Life Pictures/Getty Images
Chapter 4: "Electrodes are implanted in the brain of a schizophrenic"
by John Loengard © Time & Life Pictures/Getty Images
Chapter 5: "Scientist Theodore Tahmisian observing a grasshopper colony"
by Al Fenn © Time & Life Pictures/Getty Images
Chapter 6: "Rabies sign, England, 1989"
© Science & Society Picture Library via Getty Images
Chapter 7: "Medicine eye" by Tony Linck © Time & Life Pictures/Getty Images

ISBN: 978-1-85168-944-6
Ebook ISBN: 978-1-78074-099-7

Cover design by Matt Lehman
Typeset by Tetragon, London
Printed and bound by CPI Group (UK) Ltd, Croydon, CR0 4YY

Oneworld Publications
10 Bloomsbury Street, London WC1B 3SR

For Mum and Dad

who gave me everything

CONTENTS

Prologue
RECIPE FOR A ZOMBIE

YOU SEE THEM EVERY DAY, these zombies; they're all around you.

They shamble across the cinema screen on broken limbs and snatch at girls with long blonde hair. In the closeness of your home they explode in satisfying blossoms of rotting flesh at the flick of a trigger. Their scabby hands reach out at you with stiff cardboard fingers from the comic-book display stand. When you walk home at night, you catch them in silhouette, stumbling through the shadows, confused, drunk and lost. They sit slack-faced opposite you on the bus, their will ground away by the constant rasping of the parasites buried deep in their skulls. And as you walked in duty-free sandals over the soft ground of the tropics, did you not stop to see the quiet graves where infant wasps lay spring-loaded in the chests of their comatose prey?

Wait, you think you know what a zombie looks like?

Sure, you *think* you do. You've seen the movies, and committed their model organisms to memory. Perhaps, in the spirit of first principles, you rented out *White Zombie*, the 1932 horror classic that gave birth to the first cinematic adventure of the undead. Here you watched 'Murder' Legendre slip poison to Madeleine Short, leaving her apparently dead, only to awaken in a tomb as Legendre's mind-controlled slave. Less than an hour of reel time later, the spell is broken, and Short shows no evidence of being anything less than living. The white zombie of the title refers not to Short's death and resurrection but to her state of mind. Death is merely a misdirection, a convenient way for Legendre to throw Short's cuckolded husband off the scent. The conjurer's real treachery is transforming his human victim into a willing, compliant automaton.

Perhaps you like your zombie movies more visceral. The sorcerer's zombie was killed off by George A. Romero in 1968, when the young director turned his mind to the last days of the living. Borrowing heavily from Richard Matheson's doomsday novel *I Am Legend*, Romero wrote a script that replaced vampires with hungry ghouls, and set the action at the beginning of mankind's extinction instead of at the end; he called it *Night of the Flesh Eaters*. The distribution company, Walter Reade, wanted something sexier, so the title *Night of the Living Dead* was slapped onto the title sequence, instantaneously granting zombies with an entirely new physiology and epidemiology.

To some extent, Romero's zombies were the inverse of those created through Legendre's witchcraft: though they had little will of their own, no one was controlling them. Granted, they had a gruesome taste for human flesh, and the ability to break the rules of biology – these shambling corpses seemed to be suspended between life and death.

This is not a book about fictional zombies. This is a book about what happens to the zombie when it crawls off the page and out of the screen and into our world, the really *real* world. What must a zombie do to make that journey? Could we hijack another person's body and compel it to follow our every command? Could we die and come back again?

Death is supposed to be the end, an inviolable law of nature. We're distinguished from the gods by virtue of the fact that we will die and they never do – that's why we're called "mortals". Through time and across cultures, cautionary tales have been passed down about those who attempted to break this covenant. Even Orpheus, who managed to descend into the underworld to retrieve his dead wife, could only watch in horror as she was snatched away again on reaching the surface. And of course we have only to look to Mary Shelley's infamous *Frankenstein* to find a warning against the hubris of overturning such sacred laws through the wonders of modern technology. It seems that the desire to control the body and the mind lives large in our imaginations, but what taboo hasn't been tested?

How to Make a Zombie is filled with true tales that will keep you awake at night. Our journey, if you are prepared to come with me, starts in the fetid heat of the Caribbean cane fields, where witch doctors grind skull bones into poison and shadowy bogeymen snatch children who stay out after dark. These sorcerers corrupted the fantasy of an eternal soul that lives on without a body, creating instead a monster whose physical flesh lives on without a soul. We'll watch as secret societies stab deep into the human psyche to prick two of our greatest fascinations: our desire to cheat death, and the fear of losing our humanity. From there we fly to the bitter snow-blown streets of Moscow and the gaslit rooms of London's surgeries where necromancers build machines that can breathe life into the dead. They are challenged in this race by American scientists who hope to build a master race with their own brand of immortality techniques. On the shores of a quiet alpine lake in Switzerland, a physician slices gently into the brains of his disturbed patients, while colleagues half a world away attempt to stitch the psychic wounds of unhappy people with gossamer wire and electricity, and gangsters in the jungles of Colombia prepare for their next furtive heist by ripping leaves from a borrachero tree. All of them hope to gain purchase on the brain – some to treat, others to take control. We'll pass along the black market trade routes that course from Eastern Europe to South Korea, a global conveyor belt on which the dead are disassembled and fashioned into parts for the living. And across the world, in the air and underfoot, we'll be pestered by invisible armies of assassins – bugs, worms and fungi that press into the flesh of unsuspecting victims and whisper incredible commands from their new homes.

In the course of this journey, we'll look at what it means to be human, what it means to be alive and what it means to be the master of your own destiny. The real-life zombie has much to teach us about these things. What you learn will be both arsenal and armour against those who might someday try to zombify you.

HOW TO MAKE A ZOMBIE

1
DEAD MEN WORKING THE FIELDS

> No one dared stop them, for they were
> corpses walking in the sunlight.
>
> William Seabrook, *The Magic Island* (1929)

SO YOU WANT TO MAKE A ZOMBIE? Well then, we best start at the beginning. The zombie first shuffled its way into the global consciousness in 1889, not all that long ago. It was then that *Harper's Magazine* correspondent Lafcadio Hearn ventured on an extended holiday to unearth the truth surrounding rumours that the walking dead haunted the islands of the Caribbean.

Hearn was no hack. He was an accomplished and respected journalist who'd spent a decade in New Orleans chronicling the city's people and unique cultural life. He was not shy of penning hard-nosed editorials on crime, corruption and politics. But he also had a flair for imagery, a taste for the exotic and a keen interest in folklore – he was an amateur anthropologist in many ways, and a romantic. As he explored the islands, he found himself in strange territory, where even the darkness seemed to be alive:

> Night in all countries brings with it vaguenesses and illusions which terrify certain imaginations; –but in the tropics it produces effects peculiarly impressive and peculiarly sinister. Shapes of vegetation that startle even while the sun shines upon them assume, after his setting, a grimness, –a grotesquery, –a suggestiveness for which there is no name… In the North a tree is simply a tree; –here it is a personality that makes itself felt; it has a vague physiognomy, an indefinable Me. From the high woods, as the moon mounts, fantastic darknesses descend into the

roads, –black distortions, mockeries, bad dreams, –an
endless procession of goblins. Least startling are the
shadows flung down by the various forms of palm,
because instantly recognizable; –yet these take the
semblance of giant fingers opening and closing over
the way, or a black crawling of unutterable spiders…

The local residents were not afraid of such physical darkness – the terrors that haunted them were found in the spiritual
darkness of witchcraft, Hearn said.

In Martinique, Hearn rented a room in a small mountain cot-
tage from an elderly woman and employed the landlady's son Yébé
as his guide. The previous night, the woman's daughter Adou had
refused to go via a cemetery on an errand, claiming that the dead
would have prevented her from leaving the graveyard if she had
entered it. Hearn was intrigued. Were these dead people the zombies he had heard about? No, Adou replied, the *moun-mo* could
not leave the graveyard except on the night of All Souls, when
they travelled home. A zombie, by comparison, could appear
anywhere, at any time. Adou's usually cheerful face fell into a
sombre expression. She had never seen a zombie, she whispered to
Hearn, and she didn't want to either. Asked to describe a zombie,
she answered vaguely that a zombie was one that made disorder
at night; a zombie was a fourteen-foot-tall woman appearing in
your bedroom; a giant dog that crept into your house.

Sensing that Hearn was unsatisfied with these answers, Adou
called to her mother, who was preparing the evening meal over
a charcoal stove outside. Hearn put the question to the older
woman: what is a zombie? It was a three-legged horse passing
you in the road, the old woman replied. If you were to walk

along the high road at night, and see a great fire that receded into the distance as you approached it: the zombies made those. These were *mauvai difé* – evil fires – and unwary travellers who followed them, mistaking the light for that of a nearby village, fell into deadly chasms. Even in the middle of the day, those wandering the deserted boulevards in the townships might come face-to-face with a zombie.

Suddenly Adou remembered the story of Baidaux, a harmless simpleton, which she recounted for Hearn. Baidaux lived in St Pierre with his sister, who looked after him. One day Baidaux abruptly told his sister: "I have a child, ah, you never saw it!" The sister ignored his foolish talk, but Baidaux persisted. Every day for months, for years, he told her the same thing, no matter how much his sister shouted at him to stop. Then one evening Baidaux left the house and returned at midnight leading a small black child by the hand. "Every day I have been telling you I had a child, you would not believe me," he told his sister, "Very well, *look at him!*" She looked, and saw that the child was growing taller and taller, right before her eyes. She threw open the shutters and screamed to her neighbours for help. The towering child turned to Baidaux and told him, "You are lucky that you are mad!" When the neighbours came running in, they found nothing; the zombie had vanished. This, Adou insisted to Hearn, was the absolute truth – "*Ce zhistouè veritabe!*" Though Hearn collected many tales of this sort from island villagers, he never observed a zombie in real life.*

---

* Shortly after his article was published, *Harper's Magazine* sent Hearn to Japan. He fell in love with the country and spent the rest of his life there, marrying the daughter of a local samurai and taking the name Koizumi Yakumo. He never returned to the Caribbean.

At the time of Hearn's writing, New Orleans was known as the "Gateway to the Tropics", and that sense of 'otherness' only increased the further one travelled into the Caribbean. Haiti, in particular, possessed a horrifying and intoxicating hold on white America, a slice of Old Africa laid at its feet, a land that conjured up spectres of violence, magic and mystery. Haiti was a fiercely independent nation whose slave inhabitants had risen up in 1804 and overthrown their French masters, and repelled several subsequent colonial efforts. That independence was sorely curtailed in 1915 when civil unrest jeopardized American business interests – in particular, those of the Haitian-American Sugar Company (Hasco) – and threatened to usher in an anti-US government. The US invaded Haiti and established an occupation that would last until 1934, with enduring consequences for the country and its people.

Culturally, even that most powerful of Western settlers, Christianity, has had limited success in Haiti. Though the Catholic Church was adopted as the national religion in the 1804 constitution, no amount of milk poured into the cultural mélange by zealous missionaries could cover the spice of the indigenous Taínos and imported African gods, and in the centuries before the revolt, a new religion, Vodou, had arisen.* Haitian Vodou was a mix of spiritual rites and traditions, just as the country's population was a mix of the indigenous Taínos people, the slaves brought over in their millions from Africa, and their European colonizers and captors. Vodou grafted together elements of the

---

* This spelling is typically used by anthropologists to disambiguate the religious and cultural beliefs of actual Haitians from those imagined by Hollywood scriptwriters.

native Taínos beliefs with those of the Fon and Ewe people of West Africa and Roman Catholicism. But it was also more than a religion, encompassing a complex system of narratives, deities and practices that varied from village to village. Indeed, it is said that Haiti remains "80 percent Catholic, 100 percent Vodou".

Under the tenets of Vodou a person is composed of several parts, an amalgam of human matter as complex as the religion itself. Above all, there is the *z'etoile*, or guiding star, the celestial body that steers a person's fortunes. The *corps cadavre* is literally the physical body, while the *nanm* is the spirit of the living flesh, the vitalist energy that prevents a body from decaying in that bothersome way that dead things will. The soul too is made up of parts. There is the *gwo-bon anj* (great good angel), the animating principle of the human, the will that motivates us in our actions; also the *ti-bon anj* (little good angel), which embodies our memories and awareness. A *bokor*, or sorcerer, may be able to capture the *ti-bon anj* soon after a person's death, before it has strayed too far from the body, or else draw it from a person with magic, leaving the victim apparently dead. The *ti-bon anj* is imprisoned in an earthenware jar, which becomes the *zombie astral*, while the body it leaves behind – a physical entity that is living but has no will of its own – is the *zombie cadavre*.

It's no wonder that Hearn had trouble understanding what a zombie was.

PURSUIT OF THE FLESH

Despite Hearn's purposeful globetrotting, he never managed to meet a real-life zombie. That honour fell instead to the colourful American writer and explorer William Seabrook, a member of the 'Lost Generation' of artists living in postwar Paris. "Lusty, restless, red-haired" Seabrook was an inimitable character whose incredible life was reflected in the incredible stories he wired home to *Vanity Fair* and *Reader's Digest* magazines. Seabrook spent his life indulging in every act his body could tolerate. An alcoholic and an abusive partner, he had a penchant for sadism, and it was said that he never travelled without a trusty case full of whips and chains.* Around the same time the US was invading Haiti, Seabrook volunteered for the French military, and was gassed at Verdun, earning himself a *Croix de guerre* for his heroism.

Like Hearn, Seabrook was attracted to the sensational, and travelled to West Africa to live with the Guere tribe whose members practised cannibalism, purportedly to write a book on the subject. He was frustrated by the tribal chief's inability to describe the taste of human flesh, and refused to publish his book without securing this essential detail. When he arrived back at his base in Paris, he bribed a mortuary attendant to supply him with a sample of the elusive fare. Arriving at a friend's house, he asked the cook to prepare several dishes from the strange meat, claiming it was a rare type of wild game. "The roast, from which I cut and ate a central slice, was tender," he later wrote, "and in colour, texture, smell as well as taste, strengthened my certainty

---

* The writer of Seabrook's obituary noted with some glee that Seabrook "liked to see a girl loaded with bracelets, bangles, anklets. He preferred to see her manacled." He was actually a friend of Seabrook's – as was Aleister Crowley, who noted "The swine dog W.B. Seabrook has killed himself at last, after months of agonized slavery to his final wife." Goodness only knows what Seabrook's enemies would have said.

that of all the meats we habitually know, veal is the one meat to which this meat is accurately comparable."

The adventurer's fascinations settled on the occult after the notorious English divine Aleister Crowley visited Seabrook's farm in 1919. The men drank, smoke and exchanged stories, particularly about witchcraft, over the course of a week. Seabrook was bitten by an insatiable thirst for all things related to the dark arts, and he toured the world tracking down samples of witchcraft. The zombie in particular intrigued him. He felt that it was a creature that, unlike vampires and werewolves, appeared to have no parallel in Western culture. "The zombie, they say, is a soulless human corpse, still dead, but taken from the grave and endowed by sorcery with a mechanical semblance of life," he wrote. "It is a dead body which is made to walk and act and move as if it were alive." Seabrook couldn't wait to meet one in the flesh.

In 1928, he set off for Haiti to investigate reports of black magic there, and to see if the dead really did walk the Earth. One night during his sojourn, Seabrook and his guide, Constant Polynice, sat watching the full moon rise and discussing the demons and monsters that haunted Haiti. Polynice was a countryman but not a peasant, and though familiar with the superstitions of Haiti's populace, he was not a credulous man; he viewed these stories of demons and monsters as simply that – stories. Seabrook asked his guide what he made of the tales of the walking dead that circulated around the island. Polynice's face darkened. With consternation, the Haitian replied, "I assure you that this of which you now speak is not a matter of superstition. Alas, these things – and other evil practices connected with the dead – exist. They exist to an extent that you whites do not dream of, even though evidences are everywhere under your eyes." Why else,

Polynice asked, would even the poorest farmers build masonry tombs to house their dead, other than to protect their loved ones from this awful fate? He himself had built a family tomb next to the busy Grand Source road, so that passers-by would spot anyone trying to break into it. Even then, when his brother died, Polynice watched over the tomb for four nights, shotgun in hand, until he was sure that the body was beyond the reach of any sorcerer.

Zombies, Polynice said, were forced to work the plantations, dressed in rags. They walked with a shuffling gait, their eyes glazed over, their stare vacant like that of cattle. The zombies no longer recognized their own names, or even their own existence. Only by feeding them salt or meat could the terrible spell controlling them be broken. If salt or meat touched their lips, they would realize what they had become and flee to their graves, clawing at the dirt and crawling into their tombs, where they would find a second, final death.

Polynice recounted a well-known case, about the legendary figure of Ti Joseph, who in the spring of 1918 arrived with a tattered troupe of labourers at the Hasco offices. The company occupied a clattering, bustling industrial complex sprawled across the eastern edges of Port-au-Prince; it was filled with freight cars and machinery and its chimney belched out a column of dark smoke that towered over the area. To supply the refinery, Hasco operated several sugar cane plantations, which had produced a bumper crop that year. To reap the full profits of the bounty, the corporation's management offered a bonus to all new hands who registered at their offices. From far and wide, small bands of men and women came seeking work.

When Joseph's entourage arrived at the Hasco complex,

they stood catatonic, and not a soul replied when asked their names. They were simple people from the highlands of Morne-au-Diable, a remote part of the island, Joseph explained, and the noise of the factory scared them. They would work best if stationed deep in the plantation, away from the noise as well as other workers. Joseph was granted a gangmaster's licence and led his people to their work in the fields. But rumours swirled that he had asked to have the group separated from the others for fear that someone would recognize a long-dead family member amongst his crew. During the day, these "zombies" toiled in the plantation; at night, they were fed a plain porridge, called *mayi moulin*, made without meat or salt – the two ingredients that, of course, were said to awaken zombies from their daze. Each Saturday, Hasco paid Joseph for the work carried out by his gang, indifferent to how the money was divided among the hands – or if it was divided at all.

Then one day Joseph left the zombies in the charge of his wife, Croyance, so that he could go to Port-au-Prince and enjoy the Fête Dieu carnival. Bored and taking pity on the dumb brutes, Croyance decided to take them to the town of Croix-des-Bouquets, on the northern outskirts of the capital, so that they too could watch the parades. She tied a brightly coloured scarf around her head and set out with the zombies in tow. The streets in the town were packed with people dancing, and feasting on dried fish, cassava bread, bananas, oranges, biscuits, cakes and rum. To join the celebration, Croyance bought the zombies *tablettes pistaches* (peanuts candied in brown sugar). But unbeknown to her, the baker had salted the nuts when he prepared them. With the taste of the salt, the spell binding the zombies suddenly broke. Joseph's labourers were instantly aware of the misfortune that had overcome them, and a dreadful moan went up as the zombies began their long march back to their graves in the mountains of Morne-au-Diable. The villagers, outraged at having witnessed the horrible procession of their dead relatives, pooled their money to hire a group of mercenaries who would track down and kill the sorcerer responsible. The assassins laid in wait by the roadside, and when they spied Ti Joseph, they leapt upon him, chopping his head off with a machete.

Polynice told Seabrook that he knew of similar dead men working in the cane fields not more than two hours from his home. Captivated by Polynice's tale, Seabrook sought out other incidences of zombification on the island. He uncovered another case, from 1908, recounted in Stephen Bonsal's *The American Mediterranean*. Bonsal told of a peasant from Port-au-Prince who had been buried by the preacher of a mission church. The preacher had assisted the family in preparing the man for internment, dressing him in "grace-clothes", and so had seen his face quite clearly. On the following day, the preacher had himself closed the coffin lid before it was lowered into the ground and covered with earth. Several days later, however, a traveller on the road found a man dressed in grave clothes and tied to a tree, moaning:

> He freed the poor wretch, who soon recovered his voice but not his mind. He was subsequently identified by his wife, by the physician who had pronounced him dead, and by the clergyman. The recognition was not mutual, however. The victim recognized no one, and his days and nights were spent moaning inarticulate words no one could understand.

It seemed as though Polynice was far from alone in his fear of grave robbers, or in his vivid assurances that zombies did in fact roam the countryside. Even Seabrook's colleague Dr Antoine Villiers, of whom the author claimed there was "no clearer scientifically trained mind, no sounder pragmatic rationalist" in all of Haiti, refused to discount the zombie myth out of hand. He hinted that there might be knowledge of some kind of poison circulating amongst the *bokor* that could bring about the appearance of death, so that the victims could be abducted from their graves in the night.

Piqued by this, Seabrook arranged to visit one of the sugar plantations of which Polynice had spoken so that he could see the undead for himself. Arriving at the plantation, they could see a group of men working in the field, and climbed up the hill to greet them. Seabrook wrote:

> My first impression of the three supposed zombies, who continued dumbly at work, was that there was something about them unnatural and strange. They were plodding like brutes, like automatons. The eyes were the worst. It was not my imagination. They were in truth like the eyes of a dead man, not blind, but staring, unfocused, unseeing. The whole face, for that matter, was bad enough. It was vacant, as if there was nothing behind it. It seemed not only expressionless, but incapable of expression.

Momentarily alarmed by the possibility that the walking dead might be real, Seabrook recomposed himself and grabbed the calloused hand of one of the men. Seabrook cried out "*Bonjour, compère!*", trying to make some contact with him. But the zombie made no answer, and the gangmaster, a woman named Lamercie, grew irritated with the Westerner's presence. Lamercie pushed Seabrook away, growling that, "Negroes' affairs are not for whites". It was one of the anecdotes that made Seabrook's *The Magic Island* into an instant bestseller.

NAIL IN THE COFFIN

Seabrook's book galvanized the concept that a zombie was a person seemingly raised from the dead and set to work as a slave. It's not clear why he and Hearn garnered such different answers during their quests to find a zombie, but it may have had something to do with the occupation of Haiti. Although the "postcolonial" government made great improvements to the country's infrastructure, these roads, canals and ports were built under labour conditions ripped straight out of some of the worst codes of practice from before the revolt. Life on a work crew seemed to be one step from death's door. The zombie legend embodied the loss of autonomy felt by many Haitians under American occupation. Having seen a zombie face-to-face, Seabrook concluded that "zombies were nothing but poor, ordinary demented human beings, idiots, forced to toil in the fields".

The zombie myth would not be bound by such a grounded explanation, and a series of sensational films soon churned out of the Hollywood machine to hit cinema screens. The movies featured love triangles in which, almost without fail, a (white) woman was zombified by (black) magic. United Artists' *White Zombie*, considered to be the first zombie flick, was advertised on its release in 1932 with a suitably salacious tagline: "She was not alive… nor dead… Just a White Zombie, performing his every desire!" Its 1936 sequel, *Revolt of the Zombies*, followed the First World War recruitment of zombie soldiers, who prove their mettle when a single squad defeats an Austrian force. A flood of books and even stage plays followed, making zombies so mainstream that they were already being played for laughs in 1940's *The Ghost Breakers*. A mere twelve years after a real zombie had been unveiled (to the public's horror), the whole class of automatons could be dismissed as superstition, just a joke.

Then, on a sunny morning in the spring of 1980, a Haitian woman came face-to-face with a zombie as she went about her shopping in the marketplace – and she was rightly disturbed: the dishevelled man, eyes vacant and gait unsteady, introduced himself as Clairvius Narcisse, the brother she had buried years earlier. His death had been officially recorded by doctors at the US-run Albert Schweitzer Hospital in Deschapelles. On 30 April 1962, Narcisse had been admitted with a fever; he was also coughing up blood. As his condition deteriorated he stayed in hospital, and died three days later. A sister had identified the body and authenticated the death certificate by printing her fingerprint on it in ink. The Narcisse in the marketplace claimed that he had been poisoned, buried alive, exhumed, beaten senseless and drugged, then made to work as a slave on a sugar plantation.

Soon Narcisse's case came to the attention of Dr Lamarque Douyon, the director of the Centre de Psychiatre et de Neurologie in Port-au-Prince. Douyon had been scrutinizing the reports of zombification that had a habit of surfacing in the country, but picking the threads of truth from the tapestry of folklore had proved no easy task. The breakthrough came with Narcisse. Despite documents to the contrary, two hundred witnesses swore that the man from the marketplace was indeed whom he claimed to be. Douyon interviewed family members and friends in order to gather further details on Narcisse's past. Over the course of a meticulously constructed interview, the doctor discovered that this mysterious man could recount childhood memories and nicknames that only Clairvius could have known. It was no wonder that the Narcisse family was convinced that this was no impostor.

During the investigation Douyon asked Narcisse to describe his "death" and subsequent burial. The man said he was able to hear his relatives crying near his casket; he mentioned fragments of the conversations that had occurred at his graveside. The scar on his cheek had been inflicted by a nail driven through his coffin, he reported.

With this astonishing history recorded, Douyon contacted experts around the world in hopes of unravelling the puzzle. Douyon suspected that Narcisse has been administered some sort of poison that had invoked a death-like trance. If Narcisse's story was true, any scientists who solved it would be on the verge of making a huge medical discovery, the kind on which fame and fortune rode. It was rumoured that NASA was looking for just that sort of trance-inducing concoction, so that astronauts could be put into suspended animation to survive long trips to Mars or beyond.

Douyon's hypothesis particularly intrigued Wade Davis, a young ethnobotanist on a research team at Harvard who had received the dossier. The medical notes made at the time of Narcisse's "death" suggested that a paralytic agent might have been used to suppress body movement and function, while somehow – by Narcisse's own account – leaving the victim fully conscious. Davis had previously spent years living in the jungles of South America, studying the indigenous populations and collecting thousands of botanical specimens. He was dispatched to Haiti with instructions to uncover the zombie poison and bring samples back to the US for testing.

Initially, Davis's suspicions fell on jimson weed, a plant common to the region that belongs to the same taxonomic family – Solanaceae – as deadly nightshade and is a potent dissociative hallucinogenic. The weed takes its name from Jamestown, Virginia, where in 1676 a garrison of British soldiers were drugged with it while attempting to suppress a farmers' uprising against the colonial governor. The soldiers spent eleven days chasing feathers, pulling faces and playing childish games with one another before recovering without any recollection of the events. Pointedly, they failed to quell the rebellion, and the governor was recalled to England. Jimson weed sprouts up all over Haiti, where it is known as the "zombie cucumber". Davis established a relationship with several *bokors* and arranged to purchase samples of zombie poison to be analysed in the lab at Harvard for traces of the plant.

However, on witnessing at first hand the preparations of the zombie powders, Davis realized that jimson weed was not the key ingredient in the recipe. The *bokors* were grinding down several bits and pieces known to contain powerful toxins: the skin of the cane toad *Bufo marinus*, often used to kill pests in the sugar fields; two species of deadly pufferfish, *Diodon hystrix* and *Sphoeroides testudineus*; the itching pea *Mucuna pruriens*; and the seeds of the *tcha tcha* tree, *Albizia lebbeck*. When the powders that Wade sent back were tested on rats at the New York State Psychiatric Institute, the animals fell comatose and, after several hours, appeared dead, yet sensitive electrical equipment detected a persistent if faint heartbeat as well as brain activity. The effects of the poison lasted for twenty-four hours; after that, the rats recovered. This was the result Davis had been anticipating: the creation of a state of apparent death from which the subject might be revived.

Wade set about uncovering which of the ingredients was responsible. The effects of pufferfish poison had been well documented, since in Japan the fish is served as the dangerous delicacy *fugu* which, if not properly prepared, can kill. Every year dozens of people are hospitalized after eating *fugu*; around one in twenty die. The poison is concentrated in the fish's intestines, liver and ovaries, and contains a powerful nerve agent known as tetrodotoxin, a substance one hundred times more toxic than hydrogen cyanide (the chemical used to make the highly effective pesticide Zyklon B, which was used in the Nazi extermination camps during the Second World War). If ingested, tetrodotoxin blocks the sodium channels in nerve cells, which prevents these cells from transmitting electrical impulses. In minuscule doses it produces a tingling sensation in the extremities and a mild euphoria. This effect is one of the reasons that eating *fugu* is so popular despite its peril. But chefs must be extremely careful – too much tetrodotoxin and the numbness spreads throughout the body. Soon the unfortunate diner is unable to walk, or even sit up at the table. The throat becomes paralysed, making it impossible to speak. Eventually a coma takes hold, the person becomes completely unresponsive and the alarmed restaurant staff are calling for an ambulance. A large dose of pufferfish toxin will continue to wreak its damage until the victim cannot even draw breath; death by asphyxiation follows. The most terrifying part, however, is that the toxin does not affect the brain itself; the victim remains conscious throughout the entire ordeal of his or her last meal. Davis was sure that tetrodotoxin was the zombie culprit.

Not everyone agreed. Though tetrodotoxin shuts down the body without affecting the brain, it has not been observed to trigger the long-term catatonia associated with zombie tales, meaning that the drug was not working alone. Independent tests carried out by Takeshi Yasumoto, at Tohoku University in Japan, on two of Davis's zombie powders found only trace amounts of the pufferfish toxin, nowhere near enough to have an effect. One researcher even went so far as to brand Davis's claims as a fraud. Davis hit back, pointing out that the zombie powders he collected were made from dozens of ingredients, many containing numerous toxic chemicals that might interact with the tetrodotoxin to amplify or modify its effects. In addition, he argued that the methods used to measure the amount of tetrodotoxin in the powders could inadvertently destroy much of the key agent – an unfortunate side effect of many chemical analyses.

If tetrodotoxin was an active agent in the zombie powder, Davis said, its job could be to create an initial illusion of death, thus granting the person who had applied the poison the opportunity to abduct the "corpse" without arousing suspicion. That effect would only be temporary, with the victim soon recovering and retaining all mental faculties. He therefore proposed that a second poison might be used to maintain the abducted person in a trance-like state – perhaps a Solanaceae species such as jimson weed, his first suspect.

Davis also emphasized the importance of belief in establishing the psychological effectiveness of these drugs, an idea he developed in *The Serpent and the Rainbow*, the book he wrote about his hunt for the zombie poison. Take note, zombie-makers: according to Davis, the very fact that people feared zombification gave this drug immense power over the human mind.

THE PROFESSOR AND THE BOGEYMAN

It would be over a decade before the next international expedition arrived on "the pearl of the Caribbean" to study Haiti's undead. It was led by Roland Littlewood, a professor of anthropology and psychiatry at University College London and one-time president of the Royal Anthropological Institute. Littlewood is the author of two academic publications on zombification, a subject that has become something of a minor specialty for him.

In 1996, Britain's Channel 4 television and *National Geographic* funded an expedition to Haiti, led by Littlewood. The goal: to carry out a detailed medical examination of a zombie. Littlewood's colleague on this mission was Chavannes Douyon, brother to Lamarque, who had tracked down the case history of Clairvius Narcisse. Chavannes Douyon knew of two more cases of zombi-ism that warranted investigation.

The scientific team interviewed three suspected zombies, referred to in *The Lancet* medical journal merely as F.I., W.D. and M.M. The researchers recorded full medical histories and carried out a battery of tests to see if the still-elusive zombie could be distilled to an already established medical condition. In each case, Littlewood found little clinical evidence of anything out of the ordinary. The patients all showed some form of mental impairment, though levels of functioning varied among them. F.I., a woman, and W.D., a man, needed help with everyday tasks, including feeding, dressing and washing; in contrast, M.M., though simple-minded, could be readily engaged and was capable of caring for herself; she had even openly rejected claims that she was a zombie. F.I. was diagnosed with catatonic schizophrenia and W.D. seemed to display brain damage consistent with a period of anoxia (a lack of oxygen); M.M. was believed to be suffering from foetal alcohol syndrome. Physically and mentally there was little to connect the three patients, yet each had been branded as a zombie case.

The circumstances surrounding their alleged deaths and resurrections also varied wildly. Three years after F.I.'s death in 1976, reportedly at the hands of her jealous husband, she had been found wandering near her home by a friend and was admitted to Chavannes Douyon's psychiatric hospital, where she remained institutionalized. W.D. had been identified eighteen months after his death, but he did not closely resemble the photographs provided by the family of the man who had died, and DNA tests indicated no relation to his putative parents. M.M. had been discovered by her family thirteen years after her interment, and claimed to have been held in captivity by a *bokor* one hundred miles to the north. She and her relatives joined Littlewood and his researchers on a visit to the township where M.M. said she had been held, where she was immediately recognized by the residents as a local woman, said to be simple, who had wandered away with a group of musicians during a carnival nine months before. A woman and man appeared, claiming to be M.M.'s daughter and brother; both resembled her in physical appearance and mannerisms. A dispute then broke out between the two families, with each maintaining that M.M. belonged to them and that the other was responsible for her zombification. Later DNA tests showed that M.M. was unrelated to the family she had been living with in the south for the past nine months; the northern township was most likely her real home. So, Littlewood asked, were Haiti's famous zombies nothing more than "ordinary, demented human beings", as William Seabrook had posited decades earlier? And if so, why were they being adopted as resurrected offspring?

In Haiti, the existence of zombies is taken as a matter of fact. Zombification is enshrined in law: Article 246 of the Penal Code specifically states that poisoning with the intent to produce a death-like state will be treated no differently than murder.* It is widely accepted that zombies can be recognized by their rough, nasal voices, slumped necks and slow, ponderous walks. The placid and easily subdued zombie is an analogue to the lively and potentially aggressive madman. This distinction hints that the zombie simply may be an identity bestowed upon the low-functioning mentally ill, much in the way that Europeans for centuries attributed psychosis to possession by evil spirits.

This also explains why the fear of zombification is not of the zombies themselves – who are largely perceived as harmless – but more a fear of of becoming a zombie oneself. Families go to great lengths to ensure that their loved ones are not targeted, for example keeping watch over a tomb for days until the body begins to decompose (as Constant Polynice had), or in extreme cases separating the head from the body and swapping it with the feet. Charms are used; for example, an eyeless needle and

---

* This is frequently cited as Article 249 (which is a far more mundane statute on murder with not a mention of zombies), owing to a typo in *Time* magazine back in 1932.

thread are placed in the tomb with the body, so that should the corpse awaken it will become endlessly distracted by the impossible puzzle.

But the phenomenon of zombification is far more than a tool for making peace with mental illness: it requires not just a zombie but someone to perform the zombification, as well as a narrative to explain why the spell was cast. That's too convoluted to be a label of psychological convenience – even before the Penal Code was instituted, zombification was considered a criminal act.

In a country where ten million people are crowded into just ten thousand square miles, demand for land is high. Haiti's population has doubled since 1970, and competition for ever-diminishing parcels of land often leads to disputes. And, as it happens, in the modern cases of zombification, there is often some economic conflict involved. When Wade Davis investigated the details of Clairvius Narcisse's story, he found that the family was embroiled in a dispute over land. During Narcisse's life, he had amassed considerable wealth by refusing to support numerous children he had sired and the women who were their mothers. He also refused to co-operate with his immediate family in a plan to sell off some land, insisting he retain his share.

In a similar way, W.D.'s father had taken advantage of his literacy to register all of his family's land in his own name. As a result, he fell out with W.D.'s uncle, who eventually was convicted of W.D.'s zombification and sentenced to life in prison. And as it happens, M.M. was also a relative of W.D. – she was his aunt, the younger sister of his father. The family suspected M.M.'s zombification had been carried out in revenge for W.D.'s fate. If anyone went digging deeper for evidence of a motive, W.D.'s father had reputedly been a member of the secret police that had become renowned for spreading fear and violence against those who opposed the regime of "President for Life" François "Papa Doc" Duvalier. The Haitian secret force was officially named the Militia of National Security Volunteers, MVSN, but was given the nickname "Tonton Macoute" ("Uncle Gunnysack") after a legendary bogeyman who would snatch children who stayed out after dark – as the militia allegedly employed similar tactics against their targets. It's not difficult to imagine that W.D.'s father, if he had been a Tonton Macoute, would have earned a number of enemies during his lifetime.

Still, it seems hard to believe that in one of the poorest nations on Earth, a culture would develop where complete strangers are adopted into a family, particularly when those strangers are ill and liable to be a burden. But these cases amply show that there can be very good reasons to do so. W.D.'s "father", by taking him in, managed a victory in the family feud, by getting his brother arrested on charges of zombification and discrediting his claims to the family land. Equally, adopting a homeless stranger under the pretext that he or she is a zombified family member may serve as a kind of social penance, helping to defuse tensions brought on by inequalities of wealth in the larger community.

Comparable phenomena can be seen in other cultures around the world. In another paper on zombies, Littlewood describes relations among the Bakweri of Cameroon. Under German colonial rule in the late nineteenth century, plantations were established using external labour, and as a result the Bakweri economy declined. The people's self-confidence and morale faltered, while acts of prostitution, incidences of venereal disease and accusations of sorcery grew. In the midst of the Bakweri's increasing poverty, those who held obvious wealth were eyed with suspicion, as their individual gain was perceived to have come about through some public loss. It was believed that sorcerers would enslave their dead relatives, a nefarious means to achieve riches in rough times. Those alleged to have enlisted the undead in this way were formally cut off from society. With the advent of co-operative banana plantations in the 1950s, the economy improved and tensions disappeared – and with them, the Bakweri zombies.

So are zombies nothing more than a social construct – a sociological experiment rather than physical fact? Littlewood certainly thinks so. But there is a troubling (if promising) detail buried in his findings. During his examination of the patients, Littlewood noticed that two of the three had the same strange, circular scar over their sternum, as if a catheter had been inserted into the chest to administer some unknown substance. That spot, W.D.'s father claimed, was where the sorcerer had administered the zombifying drug.

It is an unusual injury to find in any person, let alone two people both claiming to have been poisoned and held captive as zombies. According to the *The Lancet*, the *bokors* interviewed by Littlewood and his team disavowed any knowledge of such a wound – it had nothing to do with their magic. Asked what he made of the scars, Littlewood replied, "Haven't the faintest." Still, it's something to keep in mind.

THE BIG SLEEP

When three decades after Wade Davis's investigation no substance has been found that can be conclusively called an effective zombie-making drug, you would be forgiven for thinking that the hunt has been nothing more than a wild goose chase. But Haitian property seekers and their hired *bokors* are not the only people interested in the possibility of putting a body into a state of suspended animation. In a medical emergency, doctors often *need* to make zombies.

Take Dr Peter Safar, who in 1960 found fame when he developed the simple lifesaving method of cardiopulmonary resuscitation (CPR). Safar understood that bystanders were, by definition, the first individuals on the scene in any life-threatening incident. Although different elements of the technique had been fluttering around the medical establishment for some time, he was able to corral them into a clear, workable resuscitation method that was simple enough for regular people to perform. When colleagues questioned whether exhaling into another person's lungs could fulfil their oxygen needs, Safar successfully demonstrated the technique on student volunteers who had been chemically paralysed with the drug curare.

Curare was introduced to the Western world by the adventurer and naturalist Charles Waterton, a thrill-seeking eccentric who had disappeared into the South American rainforest without shoes on his feet, determined to identify the powerful poison used by Amazonian hunters in their blow darts and arrows. Journeying up the Demerara River with six native guides, he was able to convince three tribes to share the secret of their poison. In all cases, it was made from a liquid extracted from the *urari* vine. Arriving back at his estate in England with a quiver containing several hundred poison-tipped arrows, Waterton used the agent on various animals, including an ass, demonstrating that it paralysed the voluntary muscles but did not affect the heart.

Using a bellows, he was able to keep the ass from asphyxiating, presaging Safar's experiment with students a century later.

Contrary to the heroic scenes in Hollywood movies, CPR alone is rarely enough to bring someone coughing and spluttering back to life from the jaws of death. Rather, the goal is to keep the brain supplied with oxygen until a paramedic arrives on the scene. If the brain's oxygen supply dips below its oxygen demand, damage begins almost immediately. Consciousness is lost within seconds, and death occurs minutes later. Mouth-to-mouth resuscitation sustains the patient's oxygen levels at a minimal level, and steady chest compressions carry that oxygen around the body in the blood, maintaining the brain's oxygen levels. This method saves untold lives throughout the world, even if only by purchasing a little extra time for the victim until full medical attention is available.

In the event of natural disasters and battlefield injuries, however, such professional help might be several hours away. There is no defined limit on how long CPR can be performed – efforts stretching over an hour are not unheard of. Still, it's exhausting work, even for the fittest first aider, and there's no guarantee that the patient will respond to later treatment. That's why some maverick thinkers began to ask, *What if, instead of maintaining the brain's oxygen supply, you were able to reduce its demand drastically?* In other words, what if you could turn off the person's brain for a while?

Normally the body has an operating temperature of around 37°C (98.6°F). Too high and you've got a fever, and eventually, heat death; too low and you flirt with hypothermia, the metabolic activity in your cells is suppressed and your brain grows groggy until you eventually fall unconscious. If you're not warmed up soon after that, you're destined to freeze to death. But interestingly, as metabolic activity is reduced, so too is oxygen demand. A person in a hypothermic state requires just a fraction of the oxygen that a healthy human consumes. For every degree drop in core body temperature, the metabolic rate of your brain decreases by 3 to 5 percent. When this is done on purpose, the doctors call it "therapeutic cooling".

Therapeutic cooling is now standard practice in emergency care in the treatment of drowning, strokes, cardiac arrest and blood loss – any injury that deprives the brain of oxygen. Patients are draped in wet blankets and infused with chilled saline; alternatively a special refrigerated catheter is inserted into the femoral artery, the main artery running through the thigh, to cool the blood running past it. According to a 2009 study carried out across thirty hospitals in six countries, infants are particularly responsive to the treatment. In the study, more than three hundred newborns at risk of injury following problematic deliveries were placed not in warm cots but on specially designed cooling mats that reduced their body temperature by about four degrees. This seems contrary to how we should treat such a tiny, fragile thing, and yet the cooling was a huge success: they observed a 57 percent increase in the number of infants surviving without any sign of brain damage. The tactic even seems to work when used some time after a period of asphyxiation.

The state of the science is changing every year. Over the past few years, a San Diego-based company called BeneChill has developed RhinoChill, a portable cooling system for paramedics to use on people (not rhinoceroses). The RhinoChill system squirts a blast of perfluorocarbon coolant into the nose (*rhino* in Greek) and, as the fluid evaporates off the nasal membranes, it cools the fine blood vessels running beneath them, helping to chill the brain. Normally, the act of breathing sends air rushing over these membranes for the same purpose. RhinoChill turbocharges that effect.

If lowering a person's core temperature by a few degrees can suppress brain function and allow them to survive a few extra minutes without oxygen, how far can you push it? Scientists at the Safar Center for Resuscitation Research at the University of Pittsburgh decided to investigate that question. The team experimented with states of extreme hypothermia, way beyond those ever seen in hospitals, in the hopes of achieving a true state of suspended animation, a body so cold that the metabolic rate drops nearly to nothing. Using dogs as test subjects, the researchers set out to push body temperature below 10°C (50°F). Attempting this degree of cooling with wet blankets would take too long, risking brain damage to the animals as they gradually succumbed to hypothermia. Instead, extreme hypothermia was induced by draining blood from the dogs and replacing it with ice-cold saline, using the dogs' circulation system as a heat exchange. In this state, the dogs showed no signs of life: their hearts lay still and their lungs did not draw the slightest breath. Sensors indicated no brain activity. They were, by all normal standards, dead. But, after replacing the lost blood, the dogs could be warmed enough that an electric shock from a defibrillating unit would get their hearts beating again. The science was far from perfect – some dogs showed signs of neurological injury, exhibiting physical and behavioural problems, though many survived the experiment with no lasting damage. (We'll hear more tales of undead dogs in the next chapter.)

The institute's director, Dr Patrick Kochanek, said his team hoped to push past the four-hour boundary. Asked if he was creating the equivalent of zombie dogs, Kochanek became somewhat testy. "It's very scientific and those types of words are totally inappropriate. This is an attempt to buy a little time for people who would otherwise just die", he said. Instead, the doctor prefers the term "controlled death state", the key being that the animals are clinically dead, but they are not dying.

Peter Safar, for his part, continually emphasized that he was not trying to cheat death with CPR. One of his favourite coinages, listed as rule no. 20 in "Peter's Laws for the Navigation of Life", was "Death is not the enemy, but occasionally needs help with the timing". In emergency situations, the fight to keep the patient alive can prove difficult and dangerous. CPR was merely an insurance against untimely death, helping to make sure that circumstances did not make a survivable injury a mortal one. Safar's techniques were meant to seize a second chance for those struck down by accident and misfortune, not prevent death itself. Perhaps, as the Safar Center's extreme-temperature research hints, the best way to ensure life would be to "kill" the injured on the spot, and keep him or her preserved in a state of minimal metabolic activity until the patient's body could be patched together in a stable operating theatre. After all, a brief spell of death is preferable to a permanent one.

A DEATHLY STATE OF MIND

Therapeutic hypothermia is not all that new. Its benefits have been recognized since as far back as 400 BCE, when Hippocrates described the practice of packing wounded soldiers in snow and crushed ice. Unfortunately for the modern military, there is no guarantee that battlefields will have a ready supply of snow. To make things more difficult, the human body is 70 percent water, giving it a high specific heat capacity. In layman's terms, this means it takes a lot of energy to lower a human body's temperature. To reduce the temperature of an average-sized human by just one degree, you'll need to draw out somewhere in the region of 245 kilojoules (kJ) of energy – enough heat to boil a pot of tea for your rescuers. This makes chilling quite a slow process, unless extreme measures are taken. The human body also actively resists cooling, and for good reason. Trying to push body temperature below our intrinsic limit can be very dangerous, resulting in a dramatic increase in the risk of cardiac arrest.

It's nothing less than astonishing, therefore, that so many animals can do just that, shutting down their metabolism and entering into a state of suspended animation. When small rodents such as ground squirrels enter hibernation, their body temperature can drop to just a few degrees above freezing; one Arctic species, the spring peeper toad, can tolerate a core temperature several degrees *below* freezing, protected by a kind of biological antifreeze in its blood.* These animals' tiny hearts beat several times per minute, instead of several hundred, which is the usual state of affairs; their oxygen consumption drops by 98 percent. Yet, despite these enormous changes in physiology, the animals remain perceptive of their external environment, their body carefully regulating internal temperature and other essential life functions, standing ready to wake up fully when spring arrives. In a less extreme example, birds and other animals

---

* The spring peeper toad is in a class of its own: it actually freezes solid for most of the Arctic winter.

can undergo periods of metabolic suppression during the night, when their core temperature and oxygen consumption drop. In some environmental conditions, reptiles exhibit a similar dormancy, known as aestivation. Even some fish can do it. So why not humans?

Recently, scientists have discovered several chemicals that can indeed send warm, healthy animals such as humans into a state of suspended animation – the equivalent of the zombie drug that Wade Davis searched for in Haiti. The research came out of the Pentagon's Defense Advanced Research Projects Agency (DARPA), the US military think tank responsible for the development of technologies such as unmanned drones, stealth ships and the Internet. In 2007, reporter Noah Schachtman revealed work conducted by Mark Roth, a biochemist at the Fred Hutchinson Cancer Research Center in Seattle, who was trying to trigger metabolic flexibility in humans. Roth had investigated various compounds, such as the pufferfish poison tetrodotoxin, but with no success.

Roth knew the stories of people surviving for hours while trapped under frozen lakes, and wondered whether oxygen deprivation might be the key to activating a dormant state in humans. In zebrafish and fruit flies, reducing the available oxygen has the effect of sending the animals into suspended animation, from which they later recover without injury. Unlike the slow-burn state of hibernation, this oxygen-deprivation trick pushed the animals' heart rate and breathing to zero. The problem was that it didn't work on larger animals, and especially not humans. Then Roth happened upon the answer: hydrogen sulphide.

A colourless gas with the characteristic smell of rotting eggs, hydrogen sulphide is best known for its association with volcanic eruptions and with sewers, where it is produced by microbes breaking down organic material. In enclosed spaces, these highly toxic clouds are incredibly dangerous; people working in sewers, mines and caves can be incapacitated and killed by the gas in seconds. (Rescuers check victims' pockets for loose change, since any discolouration of the copper coins is a good indicator that hydrogen sulphide gas was responsible.) Yet, hydrogen sulphide is also an important signalling molecule in our bodies, and is produced naturally in our digestive tract, as you may have noticed. As such, the body has ways of neutralizing the chemical and can tolerate small doses of it. When too much hydrogen sulphide enters the bloodstream, these defences are overwhelmed, with the gas disrupting the body's ability to use oxygen correctly and preventing cell respiration as a result.

Roth wondered whether hydrogen sulphide could be used to control the body's reaction to oxygen deprivation. The US Defense Sciences Office thought it was worth a look and agreed to fund his research. The findings were spectacular. Mice placed in an environment containing 5 percent oxygen (compared to the Earth's usual 21 percent) fell unconscious and died within fifteen minutes – not so surprising. But a second group of mice were given a blast of hydrogen sulphide before being placed in the low-oxygen chamber. These mice survived for six hours.

Technically, mice do not hibernate, but enter a mild state of metabolic suppression known as torpor if their food supply is restricted, so that their bodies can conserve resources. Mice in a naturally occurring torpor maintain a core body temperature of 26°C (78°F) or higher, even when their surroundings drop as low as 8°C (46°F). If they get too cold while in torpor, the body increases metabolic activity to compensate for the heat loss. This careful maintenance of core temperature suggests that even during metabolic suppression, mice have functioning biological processes that demand an optimum temperature be kept. When administered a dose of hydrogen sulphide, the mice could be pushed far below this natural limit, dropping their body temperature to 15°C (59°F) without suffering ill effects. Roth's experiment implied that hydrogen sulphide could be used to safely induce hibernation in non-hibernating animals, and perhaps, one day, even induce fully fledged suspended animation in humans.

Indeed, DARPA wanted to know if animals in suspended animation could tolerate injuries that would be life-threatening – if so, the research might be viable as a method of preserving soldiers who would otherwise die while they awaited the arrival of medical evacuation helicopters. To model the type of blood loss that would kill a human soldier, Roth put mice into suspended animation using the hydrogen sulphide gas and drained 60 percent of the animals' blood – a fatal amount for a normal mouse. In their suspended state, the mice survived ten hours or more. At the 2010 TED conference, Roth announced that his lab was about to enter human trials of the hydrogen sulphide therapy, even though other research teams had failed to replicate the effect in larger mammals. Two clinical trials scheduled by Roth's company, Ikaria, were later abandoned.

Roth is not alone in this field, and neither is hydrogen sulphide the only real-world zombie drug identified. Another scientist exploring the possibility of suspending life in humans is Dr Cheng Chi Lee, an associate professor of biochemistry and molecular biology at the University of Texas Medical School. Lee was interested in probing the biological mechanisms that prompt animals to enter hibernation. He reasoned that there should be some kind of signalling molecule carried in the blood that instructed the body's cells to enter a dormant state. Eventually, Lee and his colleagues isolated a molecule called 5–prime adenosine monophosphate, or 5'-AMP. In a test carried out on rodents, during periods of torpor the levels of 5'-AMP in the rodents' blood tripled, suggesting that the molecule was responsible for triggering the state. When a synthetic version of the chemical was injected into well-fed mice, the effect was as quick as it was dramatic: within a minute the animals' heart rate and body temperature plunged to one-third of normal levels – a much deeper state of metabolic suppression than normally occurs. Somewhat accidentally, Lee's team had stumbled upon a powerful agent that disabled thermoregulation, the first natural biomolecule shown to induce severe but reversible hypothermic states. Since then, other labs have found that hydrogen sulphide and 2–deoxyglucose have a similar inhibiting effect on metabolism.

All three chemicals are being investigated for treatments such as slowing the growth of cancerous tumours and preventing damage to heart tissue during surgery. However, such breakthroughs might not be as straightforward as finding the right drug. In non-hibernating animals, like humans, short periods of metabolic suppression may be tolerable, but have never been attempted for more than a few hours. Even minimal levels of metabolic activity will build up waste in the body, products that cannot be excreted while an animal is in stasis. Creatures that hibernate tend to have a high degree of tolerance for waste-product toxins, as well as a suite of biological tricks to neutralize or store waste in the body, adaptations that we humans don't enjoy. For that reason alone, extended periods of suspended animation may not even be a possibility for us.

The state of the mind during suspended animation poses another challenge to humans. Hibernating ground squirrels, for instance, cycle through regular periods of warming, the body moving up to a higher metabolic state before cooling off again. In the past, it was believed that these cycles were necessary to carry out essential maintenance on the body that could only be achieved at higher temperatures. But the real function may be even weirder: it seems that, during hibernation, animals cannot sleep. A long period of hibernation leads to sleep deprivation, and the first thing an animal does when it rouses is catch up on its sleep, which is what the ground squirrels were doing during their short stints of rewarming. It seems that suspended animation is somehow incompatible with sleep, and may even resemble a waking state. Suddenly, a four-year voyage to Mars frozen in suspended animation sounds less like a dream trip and more like a living nightmare –*The Right Stuff* zombified.

Much as Haiti's zombies built the foul *mauvai difé*, they also stoked the bright promise of a wonder drug that has attracted curious minds the world over, then retreated from all who chased after it, leaving us in the dark. The mission to uncover a technique that put the living into a death-like state – but still kept them from the jaws of death – has offered some clues, but has not yet been fully rewarded. The living might hold on a little longer, but we can't pull people back from the grave.

Because genuine necromancy would be impossible, naturally. And surely no one would actually try to raise the dead. Or would they?

2

TIME FOR A REVIVAL

We've taken a human life! If we can't
restore that life, the law will call it murder.

Dr Henryk Savaard (Boris Karloff), *The Man
They Could Not Hang* (1939)

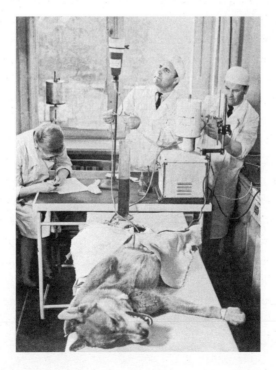

THE YEAR IS 1943. The world is at war. The Red Army has recently liberated Kiev from the Germans, and General Eisenhower has just been put in charge of the Allied Expeditionary Force to plot the invasion of Western Europe. In Manhattan, over a thousand scientists huddle into a theatre as part of the Congress of American-Soviet Friendship. There, they are screening *Experiments in the Revival of Organisms*, a twenty-minute film about the Russian scientist Sergei Bryukhonenko, who brings dead animals back to life.

The audience watches in amazement as the eminent British-born geneticist and evolutionary biologist J.B.S. Haldane narrates the proceedings. He introduces us to the spartan laboratory of the Institute of Experimental Physiology and Therapy in Moscow, rendered in murky black-and-white footage, where white-frocked men and women are attending to organs liberated from the confines of their host. Like a stranded sea creature, a glistening pair of lungs heaves and collapses in a tray as air is delivered to it via mechanical bellows. A dog's heart, suspended from tubes ferrying blood, throbs and jerks like a macabre Christmas bauble.

In impeccable clipped tones, Haldane describes how the organs can function effectively for some time after the death of the host, so long as they receive oxygenated blood through a simple relay of pumps and infusers. These services are provided by Bryukhonenko's autojektor, a whirring contraption that employs a pair of pumps and a bubble chamber. The Soviet scientist isn't content to stop with just organs, though. He wants to keep whole organisms alive. "As you can imagine," Haldane rings, "technique is *everything*."

To demonstrate his theory, the test that Bryukhonenko concocted makes the film notorious even today: using his new artificial circulation technology, he keeps alive the severed head of a dog. The hapless pooch's head twitches when poked in the eye or tickled with a feather, licks at sour citric acid daubed on its nose and mouth, squints under bright lights and looks fairly exasperated as the bench it lies upon is rapped with a hammer. Bryukhonenko's team doesn't relent, poking and rapping away to prove beyond doubt that the head is still alive and conscious.

But this is not the limit of the autojektor's power. With the aid of his primitive heart-lung machine, Bryukhonenko delivers his sensational finale: a dog is anaesthetized, and then surgeons carefully slice through a vein and drain the animal of its blood. The needles plotting vital signs in ink on paper dip lower and lower, with the moment of death coming in a dramatic flourish of the nib as the animal chokes its last breath. Then, nothing. The scientists wait poised, as if frozen in position, and for a while the only thing that moves in the laboratory are the slender hands of a stopwatch. Seven, eight, nine agonizing minutes tick by. On the crown of the tenth minute, a switch is thrown and the autojektor purrs into life. Oxygenated blood begins to flow through the corpse and the insistent rhythm of the pumps slowly provokes the heart into action. A few minutes more and the dog's head spasms, as if it is sneezing; small, steady bumps punctuate the smooth line spilling out of the respirogram. The dog's pulse and breathing grow stronger and the lines they trace more confident, until the team switches off the autojektor. The animal has been restored to life.

Experiments in the Revival of Organisms ends with the patient fully recovered, tail wagging as it is joined by several other dogs that, we are told, have all been living happily since experiencing their own resurrection. So why aren't you reviving organisms each and every day?

REPRIEVE FROM DEATH

The roots of this miraculous canine recovery lay in the closing years of the fifteenth century. The bodies of tens of thousands killed by the plague in England, France, Spain and Belgium lay damp and rotting in mass graves, and above ground mildewed crops collapsed in boggy fields, wilted by unusually cold, wet weather. The newly published *Malleus Maleficarum* laid the blame on witches. Following the instructions handed down in the papal charter *Summis desiderantes*, starving villagers built pyres to burn those responsible, and bitter winds carried the smoke from them over a rancid land. To turn a phrase, death was in the air.

Medieval Europe was dying, and so too was its spiritual and political leader, Pope Innocent VIII. In a desperate attempt to revive him, the papal physicians decided to carry out a blood transfusion. Nothing of the kind had ever been attempted. Unfortunately for the Pope, the fierce anti-intellectual stance of the Church meant that key findings on the circulatory system made by the Arab scholar Ibn al-Nafis two hundred years earlier had yet to be published in Europe, and so his physicians had no idea how to carry out the operation. They drew blood from three young boys and poured it into the pontiff's mouth. The three boys died, and so did the Pope.

It was the end for Innocent, but that was probably for the best; a new age was dawning, an age of free inquiry and science, one that the Pope would not have welcomed. In that year, 1492, Copernicus enrolled at the University of Krakow to study astronomy and maths, and Columbus set sail for the far side of the globe. Soon, the Church would be reformed and the Vatican would cede much of its secular power, concentrating instead on spiritual issues. In human hands, the great tapestry of life and death began to unravel, exposing its mysteries in the revived science of anatomy, which had lain dormant since Claudius Galen in the second century.

The celebrated Flemish anatomist Andreas Vesalius, born 1514, was one of the first to challenge Galen's long-ossified treatises. Among his many experiments with animals, Vesalius carried out vivisections in which he watched the motion of the heart and arteries fading as the subject asphyxiated. By blowing into a tube fed into the windpipe of an animal, Vesalius learned he could revive the dying subject – hence the beginnings of mouth-to-mouth resuscitation. But these sparks of insight failed to catch, and Vesalius's findings were not pursued widely.

For much of history, Galen's writings included, it was clear that those without breath or pulse were dead – or soon would be. However, while that tiny ambiguity wasn't much of a gap, Vesalius had driven a wedge into it. Slowly but surely the space between living and dead was widened enough that in 1650 a young woman named Anne Green was able to slip through.

A twenty-two-year-old servant girl from Oxfordshire, Green became romantically involved with Geoffrey Read, the grandson of her employer Sir Thomas Read. When she fell pregnant, the young cad refused to accept any responsibility for the situation. Fearing disrepute, Green hid her pregnancy. She ultimately gave birth alone, to a stillborn child. As if fate had not treated her cruelly enough already, however, the infant's body was discovered, and Green was arrested for infanticide. Being poor, young and female, the odds were very much stacked against her. The Read family, keen to avoid embarrassment, abandoned her to the mercy of the court and she was sentenced to hang.

On the day of Green's execution, a large crowd gathered at the gallows and public resentment ran high. When the ladder was kicked from under her feet, the mob set about beating her and pulling on her legs; when she was cut down thirty minutes later, they trampled her body underfoot. But these outrages did not pour forth because they hated Anne Green; far from it. Instead, these were acts of mercy, intended to hasten the young woman's death and shorten her suffering. The crowd hated the Read family for relegating Green to this fate.

As a convicted murderer, Green's body was consigned to the surgeons for dissection. When the coffin was opened, the doctors present, Sir William Petty, Dr Willis and Dr Clarke, saw a faint movement in Green's breast and decided to attempt a resuscitation. Somehow she was still alive. Once the blows had done their trick and Green seemed to be breathing, the doctors poured warm cordial into her mouth and applied tight bandages to her arms and legs to encourage blood to flow to her vital organs. To keep her warm, Green was laid in bed next to a chambermaid. The surgeons bled her, carried out an enema and employed other "diverse remedies", carefully charting her recovery.

After fourteen hours Green regained consciousness and began to speak. Her memory returned a couple of days later. She was eating solid food after two more days, and after a month she had completely recuperated from her execution. The injustice and severity of the original sentence had already stirred a tremendous groundswell of sympathy, and her resurrection was viewed by many as an act of God, a divine assertion of her innocence. It must have been a very awkward month for the Read household. Thousands of wellwishers visited the invalid, and the governor of the city issued a guard to protect her, lest the judiciary try to carry out a second attempt of the sentence. With the governor's help, a pardon was obtained, and Green retired to her hometown of Steeple Barton, riding atop her coffin in a horse-drawn cart as she made her way. She married and raised a family there, and passed away peacefully some fifteen years later.

Anne Green's case was singular in its legal particulars but it was far from unique. Just over a decade earlier, in 1549, King Edward VI, recognizing the importance of anatomy to the training of England's physicians, had decreed that all students of medicine at the University of Oxford take part in at least four dissections. The writ provoked a huge demand for bodies at a time when cadavers were already in short supply. Christians at the time believed in a literal resurrection, one that would require their physical self, and so few were willing to donate their bodies to be carved up for the advancement of this new science. The shortage continued until the signing of the Charter of Charles I, in 1626, which stipulated that the rights to the remains of all persons executed within twenty-one miles of Oxford automatically fell to the anatomical institute at the university, as executed criminals were considered to have no need for their bodies in the afterlife. The system was adopted nationally with the Murder Act of 1751, which forbid the burial of executed criminals. The implication was that these bodies would either be dissected or left to carrion at the execution grounds, an ignominy intended to deter criminal activity. Of course, the anatomy schools stood ready to collect the supply of fresh cadavers granted under the statute.[*]

At the time, the principle tool of execution in Britain was the noose, which crushed the windpipe of the victim and caused death by strangulation (standard- and long-drop hanging, where the victim's neck is broken, would not arrive until the late nineteenth century). The combination of this inefficient method of execution and the tendency to deliver victims straight to surgeons' operating tables led to a century's worth of serendipitous discoveries in the field of resuscitation. In time, these resuscitations ceased to be accidental, as physicians made deliberate efforts to raise their subjects from the dead.

In 1745 a surgeon named William Tossach presented a case to the Royal Society of London of the successful resuscitation of one James Blair, "a coal miner overcome by smoke". At that time, midwives had been using mouth-to-mouth resuscitation to revive stillborn infants, and the same technique was used to restore life to Blair. The learned men of the Royal Society, however, were not impressed, insisting waspishly that "life ends when breathing ceases". You were either dead or you weren't.

---

[*] A supply, but not nearly enough, one assumes, since thefts of corpses from graveyards by so-called resurrection men continued to be widespread, capped with the infamous case of Burke and Hare, who did not even wait for the reaper to strike but instead killed unsuspecting victims in order to sell the bodies on to surgeons.

Fortunately, other surgeons disregarded such statements and continued their attempts to resuscitate the drowned, the hanged and the suffocated. Two years later, petty criminal Patrick Redmond was revived within six hours of his "death" by hanging. When he stumbled drunk into a theatre to thank his surgeon that same evening, he caused a tremendous panic among the audience, many of whom had watched his execution earlier that day.

Throughout the eighteenth century, physicians grew more confident in their attempts to push back the boundary of death. Victims of strangulation, drowning or asphyxiation were treated to early forms of artificial ventilation – typically with the help of a pair of bellows snatched from some nearby fireplace – to reverse the cessation of breathing (the first important "marker" of death, if not the only one as posited by the Royal Society). By 1791, there were sufficient numbers of physicians trying their hand at the new resuscitative medicine that a man named Edward Coleman felt it necessary to write a dissertation on "suspended respiration". He set out what he believed to be the best practices for the lifesaving treatment, making a point of chastizing doctors for attempting to pump air into the lungs without first checking that the airway was free of obstruction – a rookie mistake.

THE SPARK OF LIFE

You might say that George Foster's life became exciting only after he died. His tale was rather mundane: his wife and child were found dead in a canal, and after suspicions fell on Foster, he was convicted of their murder. On a chilly January morning in 1803, he was led to the gallows at Newgate Prison, London, and "launched into eternity". Later, after his lifeless body was cut down, he was spirited away to the Royal College of Surgeons, where an Italian physicist named Giovanni Aldini awaited him not with a set of bellows but with some copper, zinc and brine.

Born in Bologna forty years earlier, Aldini had graduated from the university of his hometown at age twenty. He soon began work with his uncle, Luigi Galvani, who in 1771 had made an astonishing discovery: a frog's legs could be made to dance when they were connected to some metal wires. Galvani called this spark of life "animal electricity", an animating force which he believed was generated in the brain and flowed down through the nerves, supplying the muscles with power.* By the turn of the century, Aldini, now a professor, was travelling around Europe to defend his uncle's findings against his rival Alessandro Volta's bimetallic electricity theory, which claimed that the frog was somewhat incidental – it wasn't the biology but the metals that mattered. On Aldini's tours, he would connect Volta's electric piles (an early type of battery) to animals in a vivid demonstration of how electricity could flow through the body to induce muscle movement. During one of Aldini's demonstrations in London, he chose to re-create his uncle's experiments not with frogs' legs but with George Foster's newly hanged corpse.

An account of the experiment is contained in *The Newgate Calendar: The Malefactor's Bloody Register*, in keeping with its reputation as a bulletin of the most notorious incidents at the

---

* An idea derived from Franz Mesmer, who thought the whole universe was permeated by this invisible energy. A bit like midichlorians. (We'll hear more from Mesmer in chapter 4.)

prison. When the electric probes were applied to the corpse, "the jaws of the deceased criminal began to quiver, and the adjoining muscles were horribly contorted, and one eye was actually opened. In the subsequent part of the process the right hand was raised and clenched, and the legs and thighs were set in motion." Foster had been "galvanized" successfully.

To some of the audience of distinguished scientists and curious laymen, it must have seemed that the dead man was indeed returning to life; the effect was so shocking that Mr Pass, the beadle of the college, reportedly died of fright soon afterwards. Foster, however, was resolutely dead, and remained that way. His friends had made sure of this, pulling at his legs from beneath the Newgate scaffold to guarantee a quick end to his suffering. Still, the pious *Newgate Calendar* was keen to note that, should Aldini have revived Foster, the convict was liable to be executed again, as the sentence clearly proscribed that "the condemned shall hang until he be dead" – no matter how many attempts it required.

The rising number of convicts revived after their execution became a point of consternation amongst lawmakers and ethicists. If a person survived hanging, should they be executed again? Some argued that the sentence had been carried out, and the prisoner could not be expected to shoulder the consequences of sloppy workmanship on the part of the executioner. Others felt that the court clearly stated that a person was to be hanged until they were dead, and if they were not, the sentence had not been carried out. At least one surgeon, in 1752, took this dilemma into his own hands when, on leaving his operating room momentarily, he returned to find the convicted murderer and recently deceased Ewan Macdonald sitting upright on the dissection table. "Possessing more professional zeal than humanity," one account drily notes, "the surgeon took a mallet and killed Macdonald outright." That done, the dissection went ahead as planned.

Aldini was aware of the legal complications that could arise from reviving the dead and always chose to frame the goal of his experiment very carefully. Applying the electric pile to Foster's corpse was meant merely to confirm the existence of animal electricity. His object "was not to produce re-animation, but merely to obtain a practical knowledge of how far Galvanism may be employed… to revive persons under similar circumstances".

Of course, if Aldini were correct in his hypotheses, the medical implications were enormous. New techniques using electricity to treat physical complaints such as paralysis and rheumatism, as well as psychological disorders, were being developed among Aldini's converts. The excitability of the human body to animal electricity suggested this powerful stimulant could be used to resuscitate victims of drowning not just hanging.

Tales of successful resuscitations using electricity had already been reported. In 1774, a have-a-go-hero managed to bring a small girl back to life using electrical shocks. It was mid-July and the first-floor windows to three-year-old Catherine Sophia Greenhill's Soho home were thrown open against the heat and stench of London. Wandering too close to the ledge, Sophia slipped and toppled to the street below. Seemingly knocked lifeless by the fall, her parents called for the local doctor, who attempted to revive the girl without success. After twenty minutes all hope seemed lost, when a neighbour, one Mr Squires, proffered his services. Squires, something of a backroom experimenter, used a strange electrical apparatus to convey shocks across

Sophia's body, but to no effect. Then something miraculous occurred: when he passed a current across her heart, a faint pulse was seen; she sighed and began to draw shallow breaths. As if waking from a stupor, the girl vomited and remained in a concussed state for the next several days. After a week, Sophia was back to normal, "in perfect health and spirits".*

Despite being revived by a shock to the chest, it's more likely that young Sophia was suffering a coma as a result of a head injury rather than a stopped heart. Still, the story made the rounds, and soon others were attempting to apply electricity to the (near) dead. The next year Peter Christian Abildgaard, a Danish veterinarian, conducted a series of experiments on

---

* The case was reported in the annual register of the Royal Humane Society, recently established by physician William Hawes and dedicated to the revival of persons seemingly dead by accident. Hawes believed so strongly in the efficacy of resuscitation that he offered a cash reward for any victim of drowning brought to him within a reasonable time frame. His society publicized methods of resuscitation, placed lifesaving equipment along the banks of the River Thames, provided lifeguards at popular swimming locations and awarded prizes for acts of bravery in saving people from drowning or asphyxiation by noxious fumes in mines, furnaces, sewers and wells. Some of the society's advice has stood the test of time. Guidelines issued to these early first-aiders included warming the victim, rubbing the body vigorously, and performing tube-to-mouth or mouth-to-mouth resuscitation. Other methods have gone out of fashion. Beyond the old favourite, bloodletting, one treatment best left to the history books is "internal exhibition of stimulants" – in less obscured terms, a tobacco-smoke enema. Ideally, the operation was carried out with a bellows, but a smoker's pipe could suffice. It's not recorded whether you were supposed to return the pipe to its owner afterwards.

chickens in which he dispatched them with an electric shock before using another shock to revive them; Dutch-born physicist Daniel Bernoulli similarly was able to revive drowned birds with an electric zap; Prussian naturalist and explorer Alexander von Humboldt was able to revive an unconscious bird by passing an electric current along the length of its body, perhaps inspired by a trip to South America where he was himself thoroughly shocked whilst catching some electric eels. Felice Fontana of Italy, on the other hand, found that the jolts he used killed his lambs and chickens outright, or put them into a "state of irreversible petrification" – a less satisfying result.

Though Aldini was too cautious to use galvanism to revive George Foster (or even the poor frightened Mr Pass), other reanimating pioneers were willing to set aside the birds and move on to more ambitious things. Fifteen years after Aldini's experiment in London, the Scottish physician Andrew Ure decided to re-create the experiment in spectacular fashion. Ure had graduated from Glasgow University in 1801 as a medical doctor, then served briefly as an army surgeon before taking up his position as chair of natural philosophy at Anderson's Institution. Although he possessed a "combative and rancorous disposition", often falling out with other academics, Ure was a great public speaker, and was well known for encouraging working-class men and women to attend his lectures.

Ure's interest in galvanism was inspired by the work of his contemporary Dr Wilson Philip, who spent many years investigating the effects of electricity on physiology. In one such experiment, Philip severed the nerve that supplied messages from the brain to the lungs and stomach in rabbits; by attaching a battery to this broken circuitry, Philip claimed to be able to restore normal function in the animals. Ure read the results with much interest, and looked for opportunities to tinker with galvanism himself.

Around this time, he was invited to galvanize a corpse during a dissection to be conducted by the famous Dr James Jeffray, professor of anatomy at Glasgow University.* This time, the starring role in the macabre theatrics was played by Matthew Clydesdale, "a middle-sized, athletic, and extremely muscular man, about thirty years of age", executed as a murderer on 4 November 1818. After hanging from the gallows for an hour, Clydesdale's body was taken directly by the police to the anatomical theatre at the university. There, amongst the skeletons of other criminals, which hung from the ceiling, Ure and Jeffray prepared their experiment.

Their battery consisted of a series of four-inch metal plates stacked alongside one another. A healthy human could comfortably tolerate the charge generated by around eight pairs of these plates; for the distinctly unhealthy Clydesdale, Ure had arranged an incredible 270 pairs. After exposing the spinal cord and several major nerves to the air, electrodes were applied to the man's neck and one heel, causing the leg to jerk violently, and almost throwing a nearby attendant to the floor. Ure gave the corpse the appearance of "laborious breathing" by stimulating the nerves of the diaphragm; when the current was applied across the dead man's face, "every muscle in his countenance was simultaneously thrown into fearful action; rage, horror, despair,

---

* James Jeffray is perhaps best remembered for inventing the chainsaw, which he modelled after a watch chain he used to excise diseased joints from his patients.

anguish, and ghastly smiles, united their hideous expressions in the murderer's face". The audience was aghast; one man fainted, and several others bolted for the door. Undaunted, Ure and Jeffray continued to the final demonstration, stimulating the nerves of the arm, whereupon Clydesdale's finger thrust out and appeared to point accusingly at various members of the gathered crowd.

For Jeffray, this performance was simply a matter of exposing the influence of galvanic force on the human nervous system; Ure, though, was much more enamoured with the potential of this new science, which he thought might prove useful in reviving the nearly, or even the completely, dead. Ure went so far as to claim that Clydesdale's heartbeat may have returned "had it not been for the evacuation of the blood" – since, following the usual medical procedures, the corpse had been drained prior to dissection. Ure could barely temper his desire to bring back the dead, if just once:

> We are almost willing to imagine that... there is a probability that life might have been restored. This event, however little desirable with a murderer, and perhaps contrary to law, would yet have been pardonable in one instance, as it would have been highly honourable and useful to science. It is probable, when apparent death supervenes from suffocation with noxious gases, etc., and when there is no organic lesion, that a judiciously directed galvanic experiment will, if anything will, restore the activity of the vital functions.

His veiled request fell on deaf ears.*

You might think that the birth of resuscitation medicine would be of great solace to the audiences of Georgian Britain, as it promised a second chance at life should they be struck down suddenly. That's a very good marketing proposition to make to a potential zombie. However, advances in medicine over the course of the eighteenth century had the opposite effect: they fuelled fear of premature burial. The frequent and quite dramatic accounts of the dead restored to full health served to focus the public's mind on the chance that people might be interred a tad too speedily. One did not want to fall ill and wake up in a coffin. And so it was that in 1788 the *Gentleman's Journal* rallied readers to the necessity of certifying a death:

> If Nature recoils from the idea of death, with what horror must she start at the thought of death precipitated by inattention, to revive nailed up in a coffin! The brain can scarce sustain the reflection in our coolest, safest moments... If electricity proves

---

* After Ure's flirtation with galvanism, he would move on to other interests. He wrote a well-received dictionary of chemistry, then a less-celebrated book on geology in which he attempted to reconcile the existing science of the time with the biblical great flood. A staunch defender of the Industrial Revolution, he decided to apply his scientific mind to the working conditions of the age. His third book, *The Philosophy of Manufactures*, demonized trade unions and argued that factory owners were typically philanthropic in nature. The volume dismissed the notion that temperatures as high as 66°C (150°F) were harmful to employees, and claimed that any health problems found among mill and factory workers were due to an "inordinate taste for bacon".

a specific against supposed death, it will be murder
to bury a dead body without making the essay. Who
can hesitate to enforce the practice, for who would
risk being stifled in a coffin, if so facile an experiment
can prevent it?

Judging the moment of death, even with the modern equipment we have today, is a devilishly tricky thing to do. Without
sensitive electrocardiographs, early doctors had to make best use
of the diagnostic tools available to them, which weren't many.
The stethoscope was a fairly new invention, a wooden trumpet
that could be used to listen to the body's internal noises. But a
weak heart beats softly, and it is easy to miss. In fact, putrefaction
was the only indelible sign that you were dead. By the nineteenth
century, "houses of the dead" were opened where bodies could be
kept for several days until the unmistakable stench of decay was
detected; the residents sometimes had a bell tied to a finger or a
toe, so that the slightest movement would attract the attention
of the staff. If waiting in a foul-smelling warehouse for a corpse
to come to life sounds like a terrible job, pity the tongue-pullers
and nipple-pinchers – two other guardians against premature
burial – whose job it was to go from corpse to corpse to see if
anyone could be woken up.

In this context galvanism became a fad amongst physicians
in the early part of the 1800s. Doctors started to follow news of
the latest experiments, and carried special canes outfitted with
compartments for storing batteries that could be assembled
into a primitive defibrillator-type unit should an emergency
strike. But while applying a current to voluntary muscles created a violent effect, the smooth muscles of the heart appeared

much less sensitive to electricity. Without a rigorous protocol for treatment, galvanism was little more than a parlour trick. The public had mixed feelings about galvanism too. Lay people started to fear that their peaceful death might be interrupted by a too eager amateur galvanist, and some stitched labels into their clothes instructing that they should not be electrified should they fall unconscious – in effect, a precursor to today's "Do Not Resuscitate" bracelet.

Galvanism seemed to be more quackery than modern medicine. Until, that is, Sergei Bryukhonenko decided it was time, once again, to raise the dead.

THE DEAD RUSSIANS ARE COMING

Bryukhonenko, the man who kept alive the severed dogs' heads, graduated from Moscow University Medical School in 1914, just in time to be drafted into the Imperial Russian Army and bear witness to the horrors of the First World War. After the Russian revolution, he worked for several years in a large hospital, before turning to his famous experiments. At the time, the field of physiology was maturing rapidly, and Bryukhonenko decided to study the intricate workings of the organs. To do so, it was necessary to keep individual organs functioning once they had been removed from their host. In a cramped and underequipped laboratory he set himself to the task of keeping organs alive.

In May 1925, at the meeting of the Second Congress of Russian Pathologists, Bryukhonenko demonstrated the fruits of three years' labour in the lab: the original heart-lung machine that he had built for his dogs' heads. Using two electric pumps, the primitive life-support system drew exhausted blood from the head and deposited it in a glass chamber where it was warmed and oxygenated, then pumped back into the animal. In these early days, this "autojektor" was not hermetically sealed, and eventually the blood supply would coagulate and the system would fail. Nevertheless, Bryukhonenko could keep a dog's head alive for about one hundred minutes. His results were met with little fanfare, however, and failed to provoke any mention in the popular press. The following year he again demonstrated the autojektor, outlining the progress he and his colleague Sergei Chechulin had made in prolonging the lifespan of their test subjects. Again, there was no coverage.

Six months later the Soviet media finally broke the silence surrounding the device, and once they did the story gathered an unstoppable momentum. Prosaic technicians mused about how it might mean that surgeons would be able to repair a diseased heart while the machine was used to keep the patient alive; the more fanciful dreamers envisioned the birth of a full-throttled immortality engine in Bryukhonenko's lab. Public dismay mounted over the conditions under which Bryukhonenko had been forced to concoct his life-support system, and the director of the Chemical-Pharmaceutical Institute was compelled to increase the provision for Bryukhonenko's research to thirty thousand roubles. The grant came from the People's Commissariat for the Protection of Health, the highest agency responsible for medical research in the USSR.

With this funding, over the next year Bryukhonenko was able to produce five papers on various aspects of autojektor experiments. He presented these at the Congress of Soviet Physiologists in 1928 – and this time, with the full backing of the Soviet government, there was no delay in provoking a media sensation. Rumours quickly circulated on American campuses that the communist scientists had succeeded in reanimating the dead. In February 1929, a student paper at the Massachusetts Institute of Technology reported the news that Bryukhonenko and Chechulin had kept a severed dog's head alive for three and a half hours with "a queer-looking affair made of glass and rubber tubing". Within the month, *Time* magazine shared a bulletin: "Vague reports have been reaching the U.S. that Russian scientists have revivified corpses". On hearing of the invention, the playwright George Bernard Shaw quipped, "I am greatly tempted to have my head cut off so that I may continue to dictate plays and books independently of any illness, without having to dress and undress or eat or do anything at all but to produce masterpieces of dramatic art and literature."

The ability to sustain an animal using a heart-lung machine allowed for a much more mechanistic view of life. Metaphysical concepts for separating the living and the dead – such as the Catholic soul or the Vodou *nanm* – were threatened with obsolescence in the face of modern medicine. If the only difference between being alive and being dead was having a heartbeat, then wouldn't a corpse revived with a machine be alive? And why shouldn't a machine take the place of a broken heart?

Keeping a head alive was one thing; raising the dead was quite another. Bryukhonenko was not the first Russian to dedicate himself to the problem. As early as February 1902, Aleksei Aleksandrovich Kuliabko of the Physiological Laboratory of the Imperial Academy of Sciences in St Petersburg had restarted a rabbit heart that had stopped beating forty-four hours previous, and went on to repeat this procedure on animal hearts up to five days post-mortem. The next year he procured the heart of a three-month-old infant who had died from pneumonia two days earlier. Using Locke's solution – a mixture containing sodium chloride, calcium chloride, potassium chloride, sodium bicarbonate and dextrose, and designed by the British physiologist Frank Spiller Locke specifically to keep excised hearts pumping – Kuliabko was able to bring the baby's heart back to life. In 1907, he developed techniques for artificial circulation that could revive a severed fish head.* Between 1910 and 1913, another Russian, Fyodor Andreyev, succeeded in resuscitating an electrocuted dog by injecting a combination of saline and adrenaline into the bloodstream and then applying an electric shock to the heart. Andreyev would later become director of the hospital where Bryukhonenko spent his postwar years, and no doubt encouraged the young doctor to explore their common interest in reanimation.

In 1929, as Bryukhonenko was attaching dogs' heads to his autojektor, Aleksei Kuliabko set aside his outmoded fish heads and prepared his most ambitious experiment yet: a secret attempt to reanimate a human. He was joined in the experiment by the "chemico-pharmacist" Fyodor Andreyev, several assistants and a man who had passed away during surgery the day before. The team arranged the corpse on an operating table and attached a tangle of pumps to the blood vessels so that they could be pumped full of Locke's solution and adrenaline. The man's heart heaved violently in his chest, and a wet choking sound erupted from his throat like a death rattle. Kuliabko's assistants fled the

---

* An achievement that, alas, came one hundred years too soon to ride the wave of popularity spawned by the animatronic singing sensation Big Mouth Billy Bass.

room in terror. Kuliabko and Andreyev kept the man's heart beating for twenty minutes before it stopped. When news of Bryukhonenko's decapitated dogs eventually made the headlines, Andreyev could not resist hinting that the science had already moved on. He told reporters: "The principle has already been demonstrated successfully. It only remains to develop the technique for surgeons to apply practically."

Perhaps disturbed by the experiment, or wary of the public reaction that might be aroused if word leaked out of a reanimated man, Kuliabko decided to carry out his future trials on dogs, following Bryukhonenko's lead. One of Kuliabko's canine subjects showed remarkable resilience: having been poisoned and revived once, it was purportedly poisoned again and left dead for several months, before being successfully revived a second time. But Bryukhonenko had heard about Kuliabko's experiments with humans and he was ready to try his own hand at them.

He enlisted the help of the experimental surgeon Sergeo I. Spasokukotey, who had helped to engineer the network of blood banks across the Soviet Union. In 1934, showing a similar level of disregard for a person's self-determination as he had shown for the laws of nature, Bryukhonenko attempted to revive a man who had committed suicide. Just three hours after the man had hung himself, the doctor slit open an artery and a vein and connected them to the autojektor. The machine steadily drew cold dead blood from the corpse and returned it warm and rich with oxygen. For several hours the team waited, listening to the whirr of the autojektor as the dead man's body slowly warmed. Then a faint sound joined them in the room: a heartbeat. As before, a death rattle gurgled in the man's throat. The man's eyelids fluttered open; he stared at the shocked doctors crowded around him "as a man in a stupor might do". But the reanimation lasted just two minutes; the experimenters, "unbearably horrified" at what they had done, immediately switched off the pumps, allowing the patient to slip back into death. After that, Bryukhonenko left his experiments to the dogs.

THE SECOND COMING OF LAZARUS

In a laboratory not dissimilar from Bryukhonenko's, three white-frocked men stand over the tiny prone body of a fox terrier. A mask is placed over the dog's muzzle to carry nitrogen and ether, but no oxygen, into its lungs. The terrier's body convulses with this anaesthesia, then grows still; within a few minutes, it is dead. Minutes pass as the men fidget with vials and tubes, waiting for a signal. At four minutes, one of the team plunges an adrenaline-filled needle into the dog's heart. Then, the asphyxiating mask is replaced with one connected to an oxygen tank and the dog is moved to a cradle, which is rocked back and forth. A vein is intubated and a fluid consisting of canine blood, saline and heparin (an anti-clotting agent) is introduced to the terrier's body. Using a stethoscope, Dr Robert Edwin Cornish listens intently for the murmur of a heartbeat. After several minutes, he cries out in triumph, "Lazarus is alive!"

The Americans were not to be outdone by the Soviets, and Dr Cornish was their hero-in-waiting. Leanly built, with hollow eyes, sallow skin and an unruly tangle of dark hair, he was a child prodigy; he had graduated from the University of California at Berkeley in 1922 at age eighteen, and earned his PhD four years later. Cornish cultivated a broad range of interests over the years, honing lenses that could help lifeguards and divers to see more clearly underwater, writing a treatise on nutrition, experimenting in fractional distillation and marketing his own toothpaste. While at Berkeley he had been recruited to the defence of a law student accused of fraudulently adding notes to his state bar examination booklet months after the deadline: Cornish was able to demonstrate that the chemical analysis of the ink, used by the prosecution, was unsound, and warned: "conclusions regarding age of writing, as determined by this test, should be viewed with extreme suspicion". The fate of the student is not known, but Cornish's scientific mind was in fine display.

It was Cornish's interest in reviving the dead that came to define his legacy. His initial experiments in the area, conducted when he was thirty, were characteristically ambitious. On 4 February 1933, a sixty-two-year-old printer walked into the Central Emergency Hospital in San Francisco and asked to see a doctor. Thirty minutes later, he was discovered dead, having suffered cardiac failure. The dead man was put in an icebox, and four and a half hours later the sallow Dr Cornish arrived at the hospital with the aim of reanimating him. Heating pads were placed on the body, and the body itself was placed on a large teeterboard. Using the board, Cornish tipped the body up and down, sloshing the blood around and generating artificial circulation.* The attempt lasted ninety minutes. At one point,

---

* A British physician, Dr Frank C. Eve, independently developed a similar device around the same time. The shifting weight of the organs induced movement of the diaphragm, creating artificial respiration. It was especially popular for use with infants suffering asphyxia, and rocking incubators were common up until the 1960s.

the "face seemed to warm up suddenly, sparkle returned to the eyes, and soft pulsations were observed in the soft tissue between windpipe and sternum". Each movement on the teeterboard elicited more pulsations, the neck throbbing at a rate of about seventy beats per minute. (Bizarrely, neither Cornish nor the coroner overseeing the experiment had a stethoscope to hand to check if the heart was actually *beating*.) Cornish tried to apply compressions to the chest to produce artificial respiration, but the man's cold-stiffened ribs would not budge. Eventually the face cooled and the man's eyes rolled back in his skull, having lost their brief sparkle. The only definitive result of the work, in Cornish's assessment: the body failed to heat up.

Nevertheless, Cornish was buoyed by this early finding. A fortnight later the body of G.J.K., a twenty-two-year-old machinist, was plucked from the frigid waters of San Francisco Bay. The fire department gave up their resuscitation efforts after three hours, at which point Cornish stepped in at the scene. This time he brought an electric blanket and lots of heating pads to warm the body, but he found that the lack of circulation meant that he was liable to "cook" the flesh immediately adjacent to the pads unless he was able to warm the whole corpse simultaneously. Cornish persisted. The machinist's body spent four hours riding the teeterboard, but his frozen veins refused to relax and let his blood flow. Eventually rigor mortis set in, ending the attempt. Cornish noted that the stiffness developed first in those parts of the body burned by the heating pads. He never detected sign of a heartbeat.

Chastened by these two failures, Cornish realized he needed to find a more reliable and gentle method for reheating corpses. Submerging the body in warm water might be the answer he was looking for. Water posed a problem, though, since its weight would press against the body, negating the effect of the rocking teeterboard.

So on the following day, when Cornish got a third chance to handle a cadaver, he made a compromise: he plunged the body into a warm bath before strapping it, dripping wet, to the board. The subject, a surveyor in his late twenties, had been electrocuted to death a full six hours earlier, but Cornish saw that his face became flushed, indicating that blood was flowing around the body. Yet, once again, Cornish's heating blanket and pads weren't up to the task. Despite being hot enough to roast the skin of the man's knees, they failed to maintain the body's temperature. Cornish reached for his stethoscope, hoping that some whisper of a heartbeat might be heard. There was no trace. Frustrated, he put a halt to further experiments until he could reassess his strategy.

He had heard about the work of Aleksei Kuliabko and others in the Soviet Union to revive and maintain the heart in isolation of the body. If a heart, several hours dead and separated from the body, could be provoked into action by pumping Locke's solution through it, then why shouldn't the same feat work, he wondered, on a heart that was still inside the body? As early as the 1900s, it was known that increasing pressure in the arteries by mechanical action was enough to provoke a heartbeat. Thus Cornish reasoned that one or more steps in his experimental set-up were flawed, and in the spirit of science he performed a series of tests to identify which ones were at fault.

Determined not to be outdone by the cold, he moved his later experiments to a laboratory kept at a temperature no lower than 35°C (95°F), equivalent to the human body. It made for sweltering work. In July 1933, at the height of the summer, he obtained the body of a dog euthanized with gas at the city pound. Cornish connected a mercury manometer, to measure blood pressure, to the dog's femoral artery. It recorded large fluctuations when the dog was rocked on the teeterboard – proving that the movement changed blood pressure. That same day, he injected an intense blue pigment – Niagara Sky Blue 6B, for stain fans – and for good measure some of Locke's solution into the femoral vein of a euthanized sheep. After twenty-five minutes on the teeterboard, the blue dye appeared in blood samples drawn from several of the sheep's major vessels – again proving that the rocking circulated the blood. Other tests showed that the board provided artificial respiration and that the oxygen inhaled by the rocking body was then carried around the body in the blood. Each step seemed to work. But the animals did not stir from the dead.

Ultimately, Cornish gave up hope of resuscitating the dead with the use of the teeterboard alone. He considered whether he was overlooking some mystery factor. Perhaps, he thought, it was that the dormant nervous system could not spring into life without some kind of stimulation. He had read reports about the effect of "electric needles" plunged into the heart – a kind of early pacemaker – but these were untested implements. So he chose instead to use a technique with a much longer track record: jiu-jitsu.

It's time to take up martial arts.

DR IVIE'S HEALING FISTS

In the early 1900s, the accomplished martial arts teacher Kanō Jigorō had gone on tour, demonstrating Japanese combat sports around the world. He campaigned to make judo, the branch of jiu-jitsu that he founded, part of the Olympic movement, and as a result it became the first Japanese martial art to gain widespread popularity. Of particular interest to Cornish and other medical professionals was a subdivision of judo known as *katsu* or *kuatsu*.[*]

Katsu had originally been created to revive students of judo who were accidentally knocked unconscious, or killed. The skilled practitioner would deliver several sharp blows and vigorous massages at specified nerve endings, bones and joints – spots such as the upper lip, temple, clavicle, instep, neck, spine and peritoneum (a membrane lining the abdomen) – with (hopefully) dramatic consequences. According to one of the first English-language guides to the sport, *The Complete Kano Jiu-jitsu* by H. Irving Hancock and Katsukuma Higashi, published in 1935:

> Blows may be struck that cause insensibility or death. Among Occidental readers there is a notion that, because one who has been killed by a fatal blow can be brought back to life, he was not really killed after all. When a fatal jiu-jitsu blow is struck in the right way, the processes of life are mechanically stopped. It requires the prompt manipulations of [katsu] to set these vital forces at work again by mechanical means, and thus to restore life.

---

* A phrase with many translations. Here I imagine *katsu* means "wake up!", a shout that is also used in Zen Buddhism to indicate achievement of enlightenment, rather than, say, "fried pork".

After the attack, the victim is flexed and, once he or she has regained consciousness, delicately walked around the room.

Since its original development, the Japanese had extended the use of *katsu* to the resuscitation of victims of other injuries, including those involving no physical attack on the body, such as sunstroke and drowning. Others had taken note of this method. In a 1938 letter to the *British Medical Journal*, Lieutenant Colonel John Wolfram Cornwall of the Indian Medical Service extolled the potential benefits that such a system might have, and encouraged fellow doctors to investigate the adoption of *katsu* where "the usual methods" failed to work. However, the practice of *katsu* was based on in-depth anatomical knowledge, and if applied incorrectly, or to people not suffering serious injury, it could prove extremely dangerous. For that reason, the technique was usually only taught to high-ranking jiu-jitsu practitioners. But have faith, zombie-makers: despite having no formal training, Dr William Horace Ivie felt he had grasped the art well enough to apply *katsu* to one of Robert Cornish's dead sheep.

On 27 July 1933, an eighty-pound lamb was killed with nitrogen gas; the resuscitation process started thirty minutes later. After ten minutes on the teeterboard, Ivie performed *katsu* on the animal. History does not record which one of the seven *katsu* Ivie used – though surely it was not no. 6, a knuckle strike to the foot that is reserved exclusively for those incapacitated by testicular injury and which could prove a bit painful when applied to a cloven-hoofed animal. Perhaps he opted for *katsu* no. 3, a knee strike targeted between the shoulder blades? Or no. 5, scalp massage? Or no. 7, "the most important form of *katsu*", according to Hancock and Higashi, "used for the treatment of any kind of injury that causes unconsciousness or apparent death"? It involves a sharp blow to a point on the mid-spine, which in layman's terms means that Ivie would have attempted to punch the sheep to life.* In any case, his *katsu* failed, as did a post-*katsu* spell on the teeterboard.

With *katsu* no longer such a promising option, Cornish searched for some new-fangled equipment that might stimulate the nervous system, the sort of untested machinery he had once dismissed. First, he managed to land himself a highly sensitive electrocardiograph from Peralta Hospital as well as a technician, Gunhild Hansen, trained to operate it. Next, he secured a "Hymanotor", a hand-cranked pacemaker with an "electrified needle" that had been developed by the New York cardiologist Albert Hyman in 1932. Unfortunately, the unhoused electromagnet of the Hymanotor interfered with the electrocardiograph, as did touching or moving the body while it was on the teeterboard. Hansen only managed to record one post-mortem heartbeat, or at least what she thought might be a heartbeat. The Hymanotor proved useless.

To complicate things, the teeterboard had been designed for humans, and Cornish's habit of flipping sheep on their backs and then tying them to the board did not make for ideal experimental conditions. The animals were stretched unnaturally, and they were often damaged in the process. He decided to build a special teeterboard for quadrupeds, using a new pattern of restraints that could hold an animal in a state of compression rather than of extension. The sheep, mind you, having spent their life with all four feet pointing downwards were not keen to spend the last few moments of it upended. Cornish noted that

---

* Insert your own "lamb chop" joke here.

the live sheep strapped to the teeterboard experienced surges in heart rate, but felt moved to elaborate that his subjects were not uncomfortable.

Following his experiments with *katsu* and the Hymanotor, Cornish remained convinced that his methods were sound. Mechanically pumping Locke's solution though an isolated heart was enough to agitate the organ into life, he was sure of that, and it seemed likely that the process could be applied to "restore irritability of the nerves". He just could not seem to succeed in creating artificial circulation and respiration. He needed to improve his methods for stimulating the nerves – what he needed was the spark of life. "In any case, restarting the mammalian heart must be regarded as only requiring the working out of a suitable technique", he wrote, sounding much like his Soviet contemporary Andreyev.

Two years later, Cornish would discover a suitable technique.

In an unpublished report written in 1933, he proposed two lines of attack: restoring the post-mortem blood to its previous vitality and studying methods of stimulating the nervous system. By 1935, he was ready to renew his experiments along these lines. For reasons unknown, Cornish switched to dogs at this stage – the tiny fox terriers were probably easier to work with than the truculent sheep.

In an optimistic mood, he named all of his test subjects Lazarus. The fate of Lazarus I is not recorded, and Lazarus II never regained consciousness. Cornish had used his teeterboard to maintain blood flow while he performed mouth-to-snout resuscitation on Lazarus II, who had been dead for six minutes when the experiment began. The teeterboard, mouth-to-snout, and various infusions of blood, saline, heparin and adrenaline, were enough to revive the dog partially. For eight hours, it lay in a coma, "whining, panting, barking, as if ridden by nightmares". Then a blood clot formed, killing the dog.

Blood clots were a common risk during resuscitation work, and it was still not known whether this brain damage occurred due to asphyxia or because the resuscitation process itself caused clots to form. Doctors used anti-coagulants such as heparin to decrease the risk, but it was a fine balance – use too much and blood would haemorrhage through the small vessels, causing a catastrophic drop in blood pressure.

Another test subject, left dead for eight minutes, survived for only five hours. Technically, this dog should have been Lazarus III – but it's not clear if Cornish even assigned numbers to his dogs. He may have intentionally chosen not to do so, in the hope that the final, successful Lazarus would supplant all memory of the previous failures.

The first breakthrough came with the next dog, who is known as Lazarus III. For this resuscitation Cornish added a new ingredient into his lifesaving cocktail – gum arabic. Commonly used to thicken and bind ingredients in confectioneries and soft drinks, gum arabic is also used in blood substitutes to help maintain blood pressure. Lazarus III lay comatose for twelve days, but on the thirteenth day he crawled slowly along his mat. (It was around this time that the press started to take notice and the dogs were misnumbered.) Though Lazarus III managed to sit, eat, snap at flies and bark, Cornish himself was not impressed. "I am afraid that the dog will be an idiot the rest of his days", he grumbled. The bony white terrier eventually learned to stand up, but it never shook off its awkwardly slow, shuffling gait or its tendency to stare vacantly out of zombie-like eyes. The dog had been irrevocably damaged by its brief encounter with death.

Cornish's methods improved the next time round. After a five-minute death, Lazarus IV came back to life delirious but robust. The dog recovered much more quickly than the earlier Lazari, and started to sit, eat, snap and bark after four days as opposed to thirteen. This was a reanimation to celebrate.

Cornish was not shy about his achievements, and regularly invited journalists to witness his experiments in action. The press lapped up the stories, publishing breathless updates on the fate of the latest Lazarus. In 1934 a film-maker named Eugen Frenke convinced Universal Pictures to adapt the experiments for the big screen, resulting in *Life Returns*, a ropey dramatization that features documentary footage of Cornish at work. Not all of the reaction was positive, however, due to Cornish's choice of dogs as his subjects. At one point he said that he planned to switch to experimenting on pigs to avoid protests, stating bluntly: "Hogs more nearly resemble humans in their digestive and circulatory systems, and have fewer friends than dogs." Unhappy with the continuous and lurid coverage, the provost of the University of California forced Cornish to vacate his laboratory. Undaunted, the young scientist continued his experiments on Lazarus V at his home, much to the displeasure of his neighbours.

He then considered another set of candidates who could serve as experimental subjects, and who had fewer friends than dogs: executed criminals. Cornish appealed to the governors of three states – Nevada, Arizona and Colorado – for permission to try out his reanimation techniques on the bodies of those who had been executed in the gas chamber. (Cornish apparently felt persons executed in the gas chamber offered the best chance for revival, which explains why he did not apply to the authorities of his home state of California: they still favoured hanging at the time.) Each of his requests were denied. Officials were in no hurry to wade into the legal and ethical quagmire that might result if Cornish succeeded.[*]

Cornish tried again in 1947, when Thomas McMonigle, a convicted murderer on death row at San Quentin State Prison, California, wrote to him requesting the procedure be carried out "in the interest of science". San Quentin had recently bought a new-fangled gas chamber, and Cornish visited the warden, Clinton Duffy, to make his case, arguing that he had tried half a dozen times to revive victims of carbon-monoxide poisoning, but was unable to get to them quickly enough. McMonigle could receive the treatment almost immediately upon certification of his death. Duffy rebuked Cornish, explaining that it took half an hour to vent the poisonous fumes from the gas chamber, plus another half hour as a safety margin. "The only way you

---

[*]   Strangely enough, Cornish wasn't the first to ask for the chance. In a calculated gamble, the inventor of the gas chamber, Major Delos A. Turner, announced he would attempt to revive Gee Jon, scheduled to be the first prisoner to be executed by Turner's device. Turner was confident he would fail in his attempt, and wanted the world to know it. The Nevada authorities also turned down Turner's request, and everyone went away happy (except Gee Jon). Likewise, when the electric chair had arrived in New Jersey in 1908, a disquieted county physician named Frank G. Scammell announced that he would attempt to resuscitate convict John Mantasanna following his execution, claiming that it would not be the chair that ended the convict's life in these executions, but the blade of the surgeon at the subsequent autopsy. Again mindful of the difficulties that any success would create, the prison warden barred him from trying.

can do this," he told Cornish, "is to sit alongside McMonigle in the other chair, so you'll be handy." Cornish snapped, "Maybe I will at that!" before storming out. He approached the warden twice more, without success, and on 20 February 1948 a tearful McMonigle floated into the lime-green chamber "like a fish in a tank". Recalling the experience in his memoirs, Duffy wrote: "I stayed a little longer than usual that morning, but nothing went wrong, and Thomas McMonigle stayed dead. I've often wondered if he could have been brought back."

As news spread of Cornish's plans, he found plenty of willing test subjects. He reported that over fifty members of the public ("mostly single men") volunteered to be killed and raised from the dead. Some were interested in making a landmark contribution to science, others hoped for financial reward. One Kansas gentleman thought $300,000 ought to cover the hazards involved. No record of any such experiment exists; Cornish soon became occupied with other questions and abandoned reanimation work, which is our loss.

ADVENTURES IN NECRONAUTICS

If McMonigle had been brought back, what tales might he have told, and what future would be in store for him in his second life? Would he have described a wondrous tunnel of light or the righteous pain of judgement? Would he have repented his crimes or become an evangelist for Robert Cornish's reanimation project?

Consider the testimony of burglar John Smith, who earned the nickname "Half-hanged Smith" in 1705 after he swung in a noose for fifteen minutes before being restored to life by a surgeon:

> I remember a great pain caused by the weight of
> my body. My spirits were in a great uproar, pushing
> upwards; when they got into my head I saw a great
> blaze of glaring light that seemed to go out of my
> head in a flash. Then the pain went. When I was cut
> down I got such pins-and-needles pains in my head
> that I could have hanged the people who set me free.

For those who believe capital punishment is a strong deterrent to criminals, also take note that despite having his execution served, Smith would be charged twice more with burglary, each time facing a possible death sentence. (He was acquitted the first time; after his second trial, which ended in a conviction, he was transported to Australia.)

Of course, you don't need to dig as far back as 1705 to find accounts of near-death experience – the magazine racks at supermarket checkouts groan under the weight of *Amazing-But-True!* tales of people who found themselves floating above the operating table and drifting towards a bright white light. Yet, so far, few people have earnestly set out to experience death. One who did was René Daumal, a French poet and writer born in 1908.

Daumal was an exceptionally precocious young man; he was publishing poetry in prominent outlets by his late teens, and taught himself Sanskrit in order to translate Buddhist teachings into his native tongue. In the spirit of youthful rebellion, Daumal rejected the existing Surrealist art cliques and formed the Simplists with some friends, publishing a journal called *Le Grand Jeu* (The Great Game). During this time, Daumal was infatuated with Buddhist notions of transcendentalism; he tried to experience various levels of consciousness and capture them in his writings.

Most infamous among these forays was the "Fundamental Experiment", during which he repeatedly poisoned himself to near-death in order to peer at what lay beyond. At first, the sixteen-year-old tried to stay awake while falling asleep, reasoning that sleep was an analogue of death, but found he could not do it. So he dispensed with analogues. "I would put my body in as close a state as possible to physiological death, but would concentrate all my attention on staying awake and recording all that presented itself to me", he wrote. His instrument of choice was a cloth soaked in carbon tetrachloride, a highly toxic chemical from the same class of compounds as chloroform. The plan was to breathe deeply through this rag, bringing himself to the point of incapacity. If he strayed beyond that, his hand would fall from his face, allowing him to breathe freely and recover, with nothing more than a terrible headache. Writing about the "experiment" in *Powers of the Word*, he described what he saw as he tiptoed to the precipice of death:

> All that was "the world" for me in my ordinary state was still there, but it was as if someone had emptied it of its substance; it was suddenly no more than a phantasmagoria, simultaneously empty, absurd, precise and necessary. And the "world" appeared in its unreality because I had suddenly entered another, intensely more real world, instantaneous and eternal, a blazing mass of reality and evidence in which I whirled around like a moth over a flame. At that moment I felt a certainty, and this is the point at

> which the spoken word must be satisfied with merely
> hovering around the facts.

Whatever the young Daumal saw during the experience could not be put into adequate words, and so the veil of death remained opaque.

Daumal ultimately grew out of his death-baiting habit and moved on to other work. Unfortunately, the repeated high doses of carbon tetrachloride had irrevocably damaged his lungs, contributing to the early onset of tuberculosis that carried him into the higher world at the age of thirty-six.

Today, necronauts have their own magazine: the peer-reviewed *Journal of Near-Death Studies*, founded by psychologist Kenneth Ring. And there is no shortage of scientists hoping to shed some light on the subject. One such study, published in *The Lancet* in 2001, reported on 344 Dutch patients who had been successfully resuscitated. Some of the patients – 18 percent – recalled having a near-death-experience (NDE), the scientific term for reports of seeing a tunnel, a light, dead family members or friends, or some other "glimpse" of the afterlife. Eerily, patients who went on to die within thirty days of a resuscitation were significantly more likely to report an NDE, and there was little else to connect the people in the subgroup – the amount of time they spent unconscious or in cardiac arrest, their medication and their overall fear of death was no different to those patients who did not recall an NDE. The authors of the study noted that if NDEs were down to physiological causes alone (such as the brain misfiring as it became starved of oxygen), you would expect that *all* patients who had been certified as clinically dead should have reported the experience.

Another aspect of NDEs are reports of an out-of-body experience (OBE) – feeling as though you have left your own body and are watching it from above. Although most cardiac arrest survivors have no memory of their brush with death, some 10 percent develop detailed memories of the experience, verified by resuscitation staff. By rights, there should have been no activity in their brains at all, let alone the ability to record their surroundings in detail, and yet many claimed to remember floating above their own bodies, looking down on medical staff who were trying desperately to resuscitate them.

Dr Parnia of Southampton University, a physician working in intensive care, wanted to know what was happening in the minds of these near-death patients. Working with staff at twenty-five hospitals in the US and UK, Parnia devised a plan to hide images in hospital resuscitation areas, these images only being visible from the ceiling. It was ingenious, but not without its challenges: by their very nature, cardiac arrest survivors are rare, and ones who improve enough to be questioned about their cognitive and emotional experiences are even more so, and Parnia was having trouble finding enough test subjects to make his findings statistically significant. The project's last update was filed in April 2011, when it was revealed that Parnia and his research team were expanding their test into Brazilian and Indian hospitals. You may want to set up a Google alert to keep abreast of the latest developments.

ENTOMBED IN BLUE

Although the Soviet pioneers of reanimation didn't know it at the time, they had an elixir of life within their grasp – but they let it slip through their fingers during the long Cold War. It was hidden in the pale blue skin of the dead messiah in Pieter van der Werff's *Entombment of Christ*, which was one of the paintings confiscated from the Rheinsberg Palace by the Red Army in the chaos following the end of the Second World War. To find out how the elixir came to be hidden in a Flemish painting we must travel back to Darmstadt in the early part of the eighteenth century, where soon-to-be-convicted heretic and alchemist Johann Konrad Dippel was attempting to unlock the secrets of mortality.

Born in 1673 to a fourth-generation Lutheran minister, Dippel was a brilliant but adversarial man. He excelled at almost everything he put his mind to, but he had an equally powerful gift for upsetting others, and the course of his life was piloted by these two opposing talents. Following a short stint working as a preacher and lecturer in Strasbourg, Dippel took up a new calling when a pastor gave him two books on alchemy: *Experimenta* by Raymond Lully, and Wilhelm Postel's *Velamen apertum*. With signature self-assurance, convinced that with these two books and his university studies of alchemy he would be transmuting lead into gold within the year, Dippel immediately bought a house on credit. Unfortunately, his first attempts failed after eight months of preparation and, with the creditors closing in, Dippel went on the run.

He decided to concentrate his efforts on iatrochemistry, a field of alchemy devoted to finding new medicines. And what better medicine to seek than the elixir of life? He roasted deer horn in an oxygen-depleted environment, creating a type of bone char, which he then heated in a distillation apparatus to extract a foul-smelling dark-coloured liquid. Convinced this substance could cure any ailment, he marketed it as Dippel's Animal Oil. Shortly after this discovery, he went to study medicine in Holland, and eventually became physician to King Frederick I. This appointment bestowed him with quite a bit of influence and name recognition. Even though animal oils were already falling out of fashion in his day, Dippel's Animal Oil remained until the nineteenth century a regular feature of the pharmacopeia, as a nerve stimulant and anti-spasmodic.

Around the time of Dippel's deer-horn cooking he had shared a laboratory with a dye manufacturer named Heinrich Diesbach. One day, as Diesbach was preparing a batch of the red dye carmine, he found himself in need of a fixed alkali with which to complete his batch. So Diesbach borrowed some salt of tartar that Dippel had previously used to purify his wondrous elixir. Much to the dye-maker's surprise, when he added the paste to his solution a brilliant blue pigment appeared, deeper and more lustrous than any blue he had ever seen. Inadvertently, he had created one of the world's first synthetic colours, Prussian blue.

The new pigment was a phenomenal success. Aside from its striking intensity (not dissimilar from Niagara sky blue 6B, it should be noted), Prussian blue did not fade over time and was relatively inexpensive as compared to its nearest competitor, ultramarine. The artists in Paris could not get enough of it; when the colour reached Japan the artist Katsushika Hokusai used it in his famous woodblock print *The Great Wave off Kanagawa*. It is the blue found in blueprints, blue rinse and laundry blueing. And as far as we can tell, the first artist to ever incorporate Prussian blue into a painting was Pieter van der Werff, in his painting *Entombment of Christ*. That Dippel's elixir should find its way first into a painting so closely associated with the Resurrection and a theme of everlasting life is delightfully poetic.

For centuries, van der Werff's *Entombment of Christ* hung in the Sanssouci Palace in Potsdam, but it was moved to Rheinsberg Castle with the rest of the royal collection during the Second World War to escape Allied bombing raids. When Rheinsberg fell to the Soviets, the collection was taken east and displayed in Soviet galleries, right under Bryukhonenko's nose, before being returned in 1958.

Dippel's success in chemistry, and his reputation in theology, overshadowed any contribution he made to medicine. During his lifetime he published over sixty tracts on philosophy and religion, some of them inflammatory, and was imprisoned for heresy for seven years. So it's not surprising that rumours should circulate of gruesome experiments – for example, that he was attempting to transplant the soul of one cadaver into another at his residence in Darmstadt, which, almost implausibly, was Castle Frankenstein.

Do the reports of Bryukhonenko's and Kuliabko's human reanimations have any more depth than the blue pigment left behind by Dippel's Animal Oil? There's no doubt that a number of Soviet experiments in the revival of organisms did take place. The archives hold plenty of authenticated photographs of severed dog's heads lying on platters and surrounded by inquisitive onlookers, as if in some modern-day re-enactment of John the Baptist's martyrdom. Further, the invention of the "artificial heart" to maintain organs outside the human body opened the way for the development of cardiac surgery. Bryukhonenko himself was a senior figure in the Research Institute of Experimental Surgery, where the first Soviet open-heart surgery was carried out in 1957, probably using a heart-lung machine similar to the one he had developed for his dogs. In 1965 he was posthumously awarded the Lenin Prize, one of his nation's most prestigious awards, for this work. Bryukhonenko's revived heads and hearts are very much fact. Many have claimed that Bryukhonenko's film was nothing but Soviet propaganda. This is undoubtedly true; it was the time of the Second World War and the Allies were developing – and publicizing – scores of new technologies with which they hoped to win it, including devices that could save their soldiers from death on the battlefield. But that doesn't mean the movie's contents were fiction: yes, the experiments were likely staged for the cameras, but they were representative of the lab's work. And a prototype autojektor designed for humans stands on display at the Bakoulev Center for Cardiovascular Surgery in Moscow.

Given all of that, the question remains whether the autojektor could actually raise anyone from the dead. Within ninety seconds of clinical death, the human brain begins to incur critical injury; starved of oxygen and nutrients, it is the first organ to die. So the notion that Bryukhonenko could have revived an animal, dog or human, dead for hours, even days, snaps at the heels of fast-retreating credulity. The likelihood that their test subjects would not have been left irrevocably crippled by even a very brief encounter with death is just as remote – as Robert Cornish's shuffling, dead-eyed Lazarus will for ever attest.

3

MICKEY FINN AND OTHER THUGS

E'en as those bees of Trebizond,
Which, from the sunniest hours that glad
With their pure smile the gardens round
Draw venom forth that drives men mad!

Thomas Moore, *Lalla Rookh* (1817)

NEITHER THE *BOKORS* OF THE CARIBBEAN nor the physicians of the space age can bring people back from the dead quite yet, but it might be possible for them to take control of a person's mind – and isn't a living zombie just as good as a dead one? Ethnobiologist Wade Davis, although it wasn't the focus of his investigation, noted that the zombie powders he collected had included dissociative hallucinogens – drugs that made people feel detached from the world around them or even from reality itself. If such extracts and techniques do exist, surely the *bokors* are not the only ones who know about them.

History is more than ready to yield evidence for us. According to legend, the bees of Trebizond could drive men and women insane. In 401 BCE, ten thousand Greek soldiers under the command of Xenophon marched through what is now modern-day Turkey as they returned from a campaign in Persia. Looking for food, Xenophon's troops raided the hives of the bees that lived wild in the woods there. The honey was delicious and sweet, packed with sugars that should have given the soldiers the energy to continue their way home. Instead, they were struck down with fits of madness. Some were only slightly affected, stumbling around as though they were drunk. Others were less fortunate; they became so weak they could not stand in their armour. Hundreds, if not thousands, of men lay scattered across the ground, a scene reminiscent of the battlefields they had recently left behind – only there was no enemy in clear sight.

What made the bees of Trebizond such fearsome beasts? There is little to distinguish them: they're no bigger than normal bees, nor more ferocious, and their sting is no worse than that of other honeybees. In fact, the bees of Trebizond don't appear any different to normal honeybees at all, and as it turns out they're not. The bees do not cause the trouble, the honey does: while it looks and tastes like normal honey, it is highly toxic. Trebizond honey is a narcotic that causes vomiting, stupor, weakness, hallucinations, paralysis and (sometimes) death.

The legend of this merciless elixir was reinforced in 67 CE, when the Roman general Pompey the Great led his army along the southern shores of the Black Sea. The fighters native to the area laid honeycombs along the Romans' route, knowing that the men would not be able to resist sampling them. The Romans were quickly incapacitated by the poison, and as they napped they were slaughtered.

After Pompey's defeat, the ruse of baiting enemies with the poison honey became an established strategy. In 946 CE, followers of Queen Olga of Kiev attacked and killed a battalion of five thousand Russian troops after presenting them with a gift of mead – kegs filled with fermented Trebizond honey. Five hundred years later, a contingent of Tartar soldiers met a similar fate after Russian forces, perhaps mindful of their ancestors' defeat, stored casks of Trebizond mead in an "abandoned" camp.*

---

* It is worth noting that alcohol made with the honey is not always deadly. Indeed, so-called *miel fou* ("crazy honey") proved so popular that, by the Middle Ages, the region was exporting twenty-five tons annually to innkeepers, who would mix it into patrons' drinks to give some extra potency to the pint.

The bees of Trebizond are not to blame for their weaponized honey. Its effects are triggered by acetylandromedol, a toxin found in the nectar of the honeysuckle azalea, a variety of rhododendron that grows in abundance across the region. Acetylandromedol is found in many rhododendron species, and incidences of "mad honey" have been reported in the world's temperate climates where these plants grow. Similar poison honey can be produced when bees visit toxic plants such as bog rosemary, mountain laurel, redoul, doughwood and oleander; concerns have been raised that hives located near poppy fields or cannabis farms may also result in a drug-laced honey. It's for this reason that beekeepers as far back as the ancient Romans have been mindful of the types of plants growing near their hives.

Beekeepers are not the only ones who have seen the capacity to overcome the human mind. Indeed, some foods can be enough to turn an entire town mad, should that be your desire.

ROTTING BRAINS

In October 1950, the general practitioner Dr Donald Johnson and his second wife, Betty, were admitted to Warneford Psychiatric Hospital in Oxford following a sudden and dramatic deterioration in their mental health. The couple had been staying at a hotel, the Marlborough Arms Inn, in Woodstock, about eight miles away from the hospital, when they had begun to feel a growing sense of anxiety, for reasons they could not place. Donald was convinced that the phones were bugged and that his mail was being intercepted; Betty complained that a portrait on the wall of a laughing cavalier kept winking at her, very inappropriately. Soon their itching anxieties blossomed into outright paranoia.

The hotel staff grew increasingly worried for the couple, particularly the husband's "giddy turns and curious bouts of automatic talking". Donald and Betty were arrested and confined to a psychiatric ward.

Betty improved in a matter of days and was discharged, but Donald remained in "a state of mental excitement", tortured by "sexual imaginings of the bawdiest and most intimate kind". He began to suspect that he had been taken prisoner by his political enemies, communists perhaps, who planned to kill him. He then changed his mind entirely – his captors were friendly and he was undergoing some kind of clandestine training for a high-level government position. His fellow inmates at Warneford were in on the act, he decided, and they would all be sent into the field together, to Central Asia. Then he was sure he was being assigned to a diplomatic post at the United Nations; he was being groomed as a husband for Princess Margaret. During periods of relative lucidity, Donald complained that he had been poisoned, drugged with some substance that was causing him to lose his mind. No one paid any attention to the accusation, and when he voiced these concerns to his visiting wife, she hissed, "Don't say that here. They'll keep you here forever." In the no-nonsense tone of a long-suffering spouse, she added, "And do stop talking rubbish all the time." Realizing that his marriage and his doctor's practice were suffering, Donald resolved to get better, and after seven weeks of being incarcerated at Warneford, he was released with a clean bill of health.

Throughout his life, Donald Johnson remained convinced that he had been the victim of poisoning. Immediately on leaving Warneford, he travelled to London's storied Harley Street to discuss his experience with an eminent medical colleague.

The specialist admitted that Johnson's symptoms bore some resemblance to intoxication by *Datura*, opium or "Indian hemp" – drugs often used for recreational purpose. Johnson insisted that he had not taken any of the drugs willingly himself. Who had poisoned him, and for what purpose, was a mystery, however. Though he had no direct proof that one of these drugs had been responsible for his time in hospital, he became a dedicated anti-drugs campaigner. His suspicions about illicit drugs were only reinforced when the *Sunday Graphic* published a dire warning in early 1951: the threat posed to the nation's health by reefer madness was "the greatest social menace this country has known", the tabloid reported.

Then, while browsing his latest issue of *The Lancet* that August, Johnson's eye was drawn to several urgent dispatches emanating from a tiny French village called Pont-Saint-Esprit. The town's inhabitants appeared to have gone mad overnight, babbling incoherently, suffering from hallucinations and fits, some even violently attacking each another. Johnson's dread returned. These reports matched his case. The Pont-Saint-Esprit mania seemed to have affected several hundred people, resulting in four deaths and numerous cases of permanent madness. Johnson searched for further information on the mass insanity.

One month after the episode, he came across an article in the *British Medical Journal*, in which three physicians – Dr J. Gabbai at Pont-Saint-Esprit and two others working at nearby hospitals – shared the details of the poisoning. All of the affected individuals had been exposed to an unidentified toxin on the same day, 15 August, although symptoms took up to forty-eight hours to materialize. The first sign of exposure was a feeling of anxiety, after which physical complaints such as nausea, abdominal pain and diarrhoea set in. As the toxin began to work through the nervous system, the victims perceived waves of heat and cold passing over them. Their heartbeat slowed and became muffled; blood pressure often fell so low that people fainted. The body felt cool to the touch. Six days after exposure, they suffered muscle cramps, their extremities now incredibly cold. Their pupils became dilated and their skin pricked with painful attacks of pins and needles. The heartbeat was so feeble that doctors could no longer detect a pulse. Oddly, the physical reflexes of the sickened villagers were enhanced. But the most bizarre stage of the outbreak was yet to come.

In their weakened state, the victims of the poisoning found they could no longer fall asleep. This insomnia was in fact the defining characteristic of the outbreak, exhibited by more people than any other symptom. For days the villagers lay in uneasy wakefulness, sweating profusely, their soaked sheets imbued with a strange odour that both patients and doctors remarked upon. The victims began to babble incessantly, and their movements were unco-ordinated. They fidgeted in their beds, grumbling that they could feel insects crawling over their skin. Dreams, normally confined to the realm of sleep, were left homeless by this never-ending state of wakefulness, and like displaced refugees the patients' dreams began to stream into the waking world. Some of the patients began to experience terrifying hallucinations. As Dr Gabbai wrote:

> Logorrhoea, psychomotor agitation, and abso-
> lute insomnia always presaged the appearance of
> mental disorders. Towards evening visual hallucina-
> tions appeared, recalling those of alcoholism. The

particular themes were visions of animals and of flames. All these visions were fleeting and variable. In many of the patients they were followed by dreamy delirium. The delirium seemed to be systematized, with animal hallucinations and self-accusation, and it was sometimes mystical or macabre.

Medics would shout and shake the patients in order to bring them to lucidity, but the nightmares always returned seconds later. Restraining patients only increased their panic. The village postman, Leon Armunier, was in the middle of his rounds when he was struck down with visions of fire and serpents twisting around his arms. He was taken to Avignon, where he was placed in a straitjacket and confined to a room with three similarly affected teenagers who had been chained to their beds. Remembering the experience sixty years on, he told a BBC reporter: "Some of my friends tried to get out of the window. They were thrashing wildly... screaming, and the sound of the metal beds and the jumping up and down... the noise was terrible. I'd prefer to die rather than go through that again." Two villagers leapt through the window of the infirmary, trying to escape the imaginary beasts that pursued them.

These more serious cases did not manifest until some ten or twelve days after the initial reports. Of the 150 or so patients who consulted doctors, Gabbai estimated that twenty-five were struck down with significant delusions. But many victims seemed to suffer no more than a minor, short-lived digestive complaint – so minor that they did not seek medical attention.

At the time, experts blamed the epidemic at Pont-Saint-Esprit on ergot poisoning, also known as "St Anthony's fire".

Ergot (*Claviceps purpurea*) is a parasitic fungus that infects rye, and other grasses with open flowers, by forming dark powdery staves on the heads of the grain in place of the natural seed. When grain contaminated with ergot is ground into flour, the resulting loaves of bread contain toxic alkaloids that cause cramps and muscle spasms. The French call this product *pain maudit* – "cursed bread". In some individuals, ergot poisoning restricts blood flow to the extremities, causing them to become cold; in severe cases, gangrene sets in. Outbreaks tend to occur in periods of cool and wet weather.

Ergotism is one of the oldest known crop diseases, described on an Assyrian tablet dating back to 600 BCE that calls attention to "a noxious pustule on the ear of the grain". A severe outbreak appears to have struck in the Rhine Valley in 857 CE, when an archivist made note of "a Great plague of swollen blisters consumed the people by a loathsome rot, so that their limbs were loosened and fell off before death". A cursed meal, indeed. In 1676, the French botanist Denis Dodart finally made the connection between the fungus and outbreaks of St Anthony's fire. By the nineteenth century, incidents of ergotism had grown rare, since farmers could monitor their crops for fungal growth, but they still occur today. In 1975 India traced seventy-eight cases to tainted barley, which had caused delirium and hallucinations in what is known as a "convulsive" form of infection; two years later, ergot-infected oats killed forty-seven people in the Wollo region of Ethiopia. An epidemic of gangrene in Ethiopia's Arsi region in 2001 was likewise linked to ergot poisoning.

Ergot poisoning wasn't the only theory for explaining the Pont-Saint-Esprit outbreak. Some doctors proposed that it was due to mercury leeching into food and water sources from unsafe pesticide use, or that anti-fungal agents had been applied to seeds that were milled instead of being planted; others that the town's silos were infected with mould, or that illegal additives had been used to bleach the local flour. All of these theories blamed the bread. The "staff of life" had zombified them.

Donald Johnson, nagged by the reports he had read in the medical journals, rejected this explanation. He travelled to Pont-Saint-Esprit, where he interviewed affected residents as well as their doctors. He wandered the village looking for clues. The culprit, he decided, was not bread but hemp, which he found growing wild at the edges of fields in the area, possibly the off-spring of surreptitious recreational plantings or the ancestors of long-forgotten crops (hemp has been an important source of high-quality fibre for centuries). Here he had a criminal culprit ready to be charged and convicted. Johnson decided to try his sensational case in the press, telling the *Daily Mirror* in 1952: "There is no doubt that the town went mad because it was drugged with marijuana... Some of it must have been reaped with the grain."

For the rest of his career, Johnson spoke out against cannabis and other recreational drugs. His attitudes in this arena were complex – unlike many modern moralistic campaigners, he was primarily concerned with the health effects of drugs, as one might expect of a doctor, but from 1954 until 1964 he was also a politician – the Conservative Member of Parliament for Carlisle. As an MP, he lobbied against a proposed ban on the prescription of heroin to addicts seeking treatment, and for improvements in the admission protocols and management of psychiatric hospitals. He felt that pharmaceuticals were too highly regarded over traditional remedies such as Epsom salts, but equally was opposed to the British government's proposal to give faith healers positions within the National Health Service. If nothing else, Johnson was a well-meaning if obstinate character, guilty of a slightly overinflated sense of his own worth. His autobiography, spread over five volumes, includes 1967's *A Cassandra At Westminster*, a very serious and self-referential allusion to the mythical prophetess whose predictions were always ignored and always proved to be accurate.

ACID SPIES AND SECRET HIGHS

One of the toxins present in the fungal bodies of ergot – ergotamine – is a direct precursor of LSD, and it is this similarity to LSD that gives rise to another drug-mad theory regarding the Pont-Saint-Esprit outbreak. In 2009, American writer Hank Albarelli claimed that he had uncovered evidence of a clandestine LSD experiment that had been conducted not far from the town. Among the papers that Albarelli dug up there is an alleged transcription of a 1954 conversation between a CIA operative and a representative of the chemical company Sandoz Laboratories in which, after several drinks, the scientist had blurted out: "The Pont-Saint-Esprit 'secret' is that it was not the bread at all... It was not grain ergot."

At the time of the outbreak, the only facility in the world producing LSD was run by Sandoz, the company at which, in November 1938, lysergic acid diethylamide had first been synthesized. The discovery of LSD had been an accident: the young chemist Albert Hoffman had been researching the common traits of ergot and squill, which had long been used in traditional medicines. Five years later, in the midst of the Second World War, Hoffman detected the drug's potent effects.

Although Albarelli's theories failed to gain much purchase, there is little question that the CIA quickly became interested in LSD after its discovery. The agency was gripped by the fear that Soviet scientists were cooking up ways to create sleeper agents who might infiltrate US institutions, subjects so cleverly brainwashed that they would not even realize that they were spies. Determined to find ways to root out these Manchurian candidates, and to plant a few of their own, the CIA officially launched the MK-Ultra programme in 1953. Its goal was to discover and develop mind-control techniques.

In the pressure cooker of Cold War paranoia, abundant funding, total opacity and little oversight, the MK-Ultra programme built a fever dream as bizarre and nightmarish as that suffered by the residents of Pont-Saint-Esprit. The project spanned eighty institutions, including colleges, laboratories, hospitals, prisons and brothels. The CIA also enlisted members of the US Army's Special Operations Division at Fort Detrick, Maryland, to test the effects of LSD. Agents started administering the drug to members of the public, typically without informed consent, or any consent at all. One resident of a Kentucky asylum was dosed with LSD daily for six months. During Operation Midnight Climax, prostitutes working for the CIA brought clients to safe houses in New York City and San Francisco, where the unwitting men were drugged surreptitiously and watched through a two-way mirror. Agents went as far as slipping the drug into the drinks of friends and co-workers, and this led to one of the most infamous scandals of the experimental programme – the story of Frank Olson.

Olson was an army scientist working in a germ warfare laboratory. At a company retreat in Deep Creek Lake, Maryland, he was served a glass of Cointreau spiked with LSD. According to the official account of events, he had a bad reaction to the drug. He became consumed with paranoia and depression, and then suffered a complete nervous breakdown. Sent to see a psychiatrist in New York City, Olson expressed his wishes to leave germ warfare research and do something else with his life. Shortly afterwards, the father of three threw himself from the window of his tenth-floor hotel room. When the circumstances of his death were later revealed, the CIA admitted drugging Olson without his knowledge and paid his family $750,000 in an out-of-court settlement. Conspiracy theories have raged ever since over the exact course of events that took place in Olson's hotel room. Some say that he was genuinely suicidal, others that he was assassinated for his perceived lack of loyalty.

Activities under the MK-Ultra programme were reduced in 1964 and shut down completely a decade later. Evidence of the project only came to light during an investigation into illegal intelligence gathering following the Watergate affair, conducted by the specially convened Church Committee of the US Senate. In addition to LSD, MK-Ultra had looked into the efficacy of sonic blasts, sleep deprivation, sexual blackmail, hypnosis, a variety of drugs, and much more in spycraft. It's likely that the true extent of the programme will never be known, as CIA director Richard Helms ordered the destruction of all MK-Ultra materials when the programme ended – to the regret of those hoping to discover if the CIA ever found its zombie drug.

RECRUITED, BODY AND SOUL

Mind-altering drugs have been a part of human life throughout history (and before), employed by mystics to divine the future, and by shamans to make sense of the present. However, our attempts to fuse a working knowledge of drugs with specific psychiatric aims has lagged seriously behind our ability to use pharmaceuticals for their effects on the body. Ergot, for example, has been used since ancient times for its ability to induce convulsions and act as a vasoconstrictor, which makes it a valuable medication for use during childbirth and for treating post-partum bleeding. Still, no one has been able to tame the fearsome psychological effects of the fungus. The same is true for many hallucinogens – service-able for open-ended divination, but impractical as mind-control agents, truth serums, memory blockers, or zombifying agents. But, of course, that's not to say that people haven't already tried.

One theory of the English word "assassin" is that it is derived from *hashshashin*, a notorious Persian-Arab gang whose name translates as "hashish eaters". By Marco Polo's account, this band of murderers was led by Hassan-i-Sabbah, the Old Man of the Mountain, who was said to feed hashish to his new recruits before leading them into a fortress garden filled with the most wondrous things imaginable. Every type of fruit grew there; the lodges were covered in gold; and streams of milk, honey, water and wine coursed through the garden in abundance. The novices were greeted by the most beautiful women they had ever seen, all well versed in comforting men. The garden's women could play every instrument with skill and sing gloriously. To the thoroughly stoned Muslim men, the purpose-built garden appeared to be the heaven promised to them by the Qur'an. Eventually they would fall asleep and be carried out of the fortress by Hassan-i-Sabbah's associates. On waking in the outside world, cold and hungry, the recruits were convinced that their master had given them a preview of the afterlife, and now knowing what paradise awaited them there, were willing to follow his commands without any dread of death.

The Assassins, or at least one faction of Hassan-i-Sabbah's followers referred to by that name, did indeed exist. More accurately known as the Nizari Isma'ilis, they were an offshoot of the Shi'ite Fatimid caliphate. In practice, they resembled the Christian Knights Templar – a militant religious sect that sprung up to support the Catholic Church's Crusades, but later operated in secrecy.

While the promise of nirvana and the illusion of its existence, created under the influence of cannabis, would have fostered the Assassins' fearlessness and loyalty, the drug itself could only serve as a tool to cementing devotion, not an active agent of it. And, it seems, Hassan-i-Sabbah's heavenly garden, and his fondness for feeding hash cakes to his recruits, is likely no more than an entertaining fiction. Mike Jay, an author of many books on drugs and their role in society, points out a far less colourful translation of the Arabic word *hashshashin*: in the twelfth century it was a broad term meaning "outlaws" or "riff-raff".

But, as Jay remarks, that does not mean there weren't drugs that could achieve the level of control necessary for managing a murderous band. The genus known as *Datura* takes its scientific name from the Hindi *dhatura*, or "thorn apple", because of the plant's spiky round fruit. *Datura* are related to deadly nightshade, henbane and tobacco, all of which are toxic in various degrees, but *Datura* are also related to common and bountiful food crops including the potato, tomato and chilli pepper.

Datura species have been grown since antiquity for their dissociative qualities, with recreational users including the ancient Greeks and the Aztec and Navajo tribes of the Americas. (It's worth noting that the drug is physically dangerous.) The plants contain hallucinogenic alkaloids, most especially atropine and scopolamine, a powerful drug found in many tranquilizers, including those used to treat motion sickness and sea sickness. In the nineteenth century, scopolamine was mixed with morphine to induce a condition known as *dämmerschlaf* ("twilight sleep") to relieve the pain of childbirth. The goal was to create an experience that, though not free from pain, was free from the *memory* of it. Women drugged in this way had no recollection whatsoever of giving birth. As it became apparent that newborns were negatively affected by giving the drugs to their mothers during labour, the practice was phased out. No wonder that varieties of *Datura* have gained a number of evocative aliases: sacred thorn-apple, moonflower, witches' weed, devil's trumpet, hell's bells.

Others have employed the plant for more nefarious purposes. In 1994, Anne Proenza reported in the *World Press Review* that an epidemic of drug-assisted robbery was under way in Bogota. Sometime in the 1970s, gangs in Colombia had devised a means to extract the principal toxins from *borrachero*, a woody species of *Datura* that grows into small trees and whose local name translates as "intoxicator".* One extract, a white powder known locally as *burundanga*, is completely tasteless and odourless and

---

* The name comes from the Spanish *emborracharse*, literally "get drunk". The conquistadores believed that any person or animal resting in shade of the tree would be driven mad.

dissolves readily in water – handy traits if you're criminally minded. *Burundanga* could be slipped into the drink or food of an unsuspecting victim, who would come round hours or even days later, with no memory of what had happened. If the target was lucky, he or she would have lost nothing more than some valuables and the better part of a weekend. According to Proenza, between fifteen and twenty victims of *burundanga* poisoning were arriving at the emergency ward of Bogota's Kennedy Hospital every weekend.

Of course, that was if the target was lucky. Whilst drugged, a victim could be maintained in something akin to a zombified state, carrying out whatever tasks a captor commanded. As Dr Camilo Uribe of the Bogota Toxicology Clinic told Proenza:

> The victim does what he or she is told, then forgets both what happened and who the attackers were. It's a perfect form of chemical "hypnosis" which allows all sorts of crimes. Rape and sexual abuse are the most common, but it can also lead to some more serious crimes. Some people use it as a kind of truth serum like sodium pentothal, which was tried out during the Second World War.

Uribe went on to describe other incidents involving the drug. A "well-known diplomat" had disappeared from an upmarket bar in Bogota, only to appear several days later at Santiago airport with a woman he did not know and a suitcase packed with cocaine. A senator and his wife spent the entire night travelling with a gang of thieves, withdrawing huge amounts of money from various cash machines around the city. A young American woman was discovered in a state of confusion, unaware that the entire weekend had passed. Tests showed she had had sex with at least seven different men. She had no memory of the attack.

Thus far, the use of *Datura* seems to have been limited mostly to Colombia, though access to the plant does not seem to be the issue. As early as 1985, criminals had ingeniously adjusted the recipe for the drug, switching from the *Datura*'s scopolamine to benzodiazepine, a pharmaceutical sedative that can be more easily obtained. Thereafter, drugging with pure benzodiazepine also became more common. The two agents could also be blended to create a particularly potent form of *burundanga*. The toxicology department of the Institute of Forensic Science in Bogota features a display case filled with food, drink and confectioneries found to be laced with *burundanga*. The tantalizing items include chocolate, chewing gum, sweets, *aguardiente* (or "fiery water", a highly alcoholic local spirit) and even an unopened can of Coca-Cola, adulterated using a syringe jabbed through the lid.

Nevertheless, Mike Jay is sceptical of claims that *Datura* extract could be used as an everyday zombifying agent. Scopolamine and its related alkaloids were rejected as truth serums or mind-control drugs by the likes of Josef Mengele and the CIA, and for good reason. The chemical's disorientating and dissociative powers might be attractive to interrogators considering their toolkit, as might its ability to impair memory function, but those who take the drug cannot be controlled consistently. The effects, according to Jay, are "better understood in terms of disinhibition which causes people to act in ways that they later regret". That could equally be said for the *aguardiente* to which *burundanga* was sometimes added.

The fear of being drugged unwittingly is pervasive, but the lurid stories of crimes committed and suffered under the influence of secret poisons vastly outpace their incidence. Numerous chain emails have claimed that credit cards impregnated with the drug have been used to rob victims – none of which have been proved to be true. Many of the Internet forums frequented by European backpackers reverberate with cautionary "friend of a friend" tales of prostitutes offering up dope-laced nipples and the like. In the UK, the *British Journal of Criminology* reported that half of respondents to a survey said they knew someone who had been a victim of drink spiking, though police and other authorities had no evidence to support the notion of such widespread use of debilitating drugs in sexual assaults or any other crimes. The author of the study, Dr Adam Burgess, said: "Young women appear to be displacing their anxieties about the consequences of consuming what is in the bottle on to rumours of what could be put there by someone else". It seems young men do not seem to share these anxieties, according to the medical establishment.

It's natural to worry that a drug will make you lose control of your mind – much of recreational drug use is dedicated to achieving just that. People take drugs to escape from the realities of the physical world – to manipulate their sense of time, space, self, cause and effect – in other words, facts and accountability. In a comment piece for the *Guardian*, Jay argues: "Perhaps this persistent myth of mind control can best be understood by the suggestion that, as long as we continue to be confronted with actions that would otherwise force us to take an unacceptably dark view of human nature, the explanation that 'the drugs made me do it' will continue to be both proposed and accepted." We welcome the chance to allege pharmaceutical manipulation, when it means that we aren't responsible for our unsavoury actions or thoughts.

But what if you could not say "the drugs made me do it"?

PUSHER

As it happens, the greatest successes of mind control have not been achieved with drugs; they have been the work of more humble masters of human behaviour. In 1963, the famed American psychologist Stanley Milgram revealed the results of what may be psychology's most controversial experiment: that most people are highly obedient to authority figures. Volunteers were instructed to deliver electric shocks to unseen participants – actors posing as test subjects. Most of the volunteers did as they were told, and raised the voltage to increasingly dangerous levels at the experimenter's behest, despite anguished screams emanating from their unseen "victims". Milgram had been inspired by the trial of Adolf Eichmann and sought some explanation for the atrocities of the Second World War. This proved, Milgram said, that obedience to authority could be used to convince normal people to act inhumanely.

Half a century later, this human tendency to obey authority allowed one man to take control of others in a shocking series of attacks at the very heart of the American psyche. On 9 April 2004, Donna Summers, an assistant manager of a McDonald's restaurant in Mount Washington, Kentucky, received a call from a man identifying himself as a police officer. He told Summers that a customer had reported the theft of a wallet, and described an employee whom he suspected of being the culprit. He also reported that he had Summers' supervisor, Lisa Siddons, on the other line. The description matched that of eighteen-year-old Louise Ogborn, a young staff member who had agreed to work an extra shift that night in order to pick up a few more hours' pay – her mother was sick and had recently lost her job, so the girl needed as many shifts as she could get.

Ogborn was called into the manager's tiny office, where the police officer instructed Summers to perform a strip-search. The girl was made to remove all of her clothes, leaving her wearing nothing but a small dirty apron. Summers then said she needed to attend to the busy restaurant, so the caller had her choose another staff member to watch over Ogborn and stay on the phone with the officer. That employee, twenty-seven-year-old Jason Bradley, walked out in disgust when the caller told him to remove Ogborn's apron. But that didn't put a halt to the interrogation.

With Summers back on the line, the officer asked if there was someone else she could trust to watch over the suspect. They agreed to call in Ogborn's fiancé, Walker Nix Jr, who was not even an employee of the restaurant. When Nix arrived, the officer convinced him to carry out a series of increasingly bizarre demands upon Ogborn: disrobing her completely, making her jump up and down, then physically and sexually abusing her. During these incidents, Donna Summers re-entered the office several times, yet failed to put an end to the teenager's ordeal. It was as though she had been hypnotized and had lost awareness of the reality of the situation.

When yet another man was invited to guard Ogborn, he refused the caller's demands. This did not occur until four hours after the initial report of the supposed theft. It was only at this point that Summers realized that the man on the other end of the line may not be a police officer. She called Siddons and discovered that she was asleep; Summers' supervisor had never been involved in the phone call. CCTV footage of the McDonald's back office shows it filling with senior staff as they begin to comprehend the enormity of what has just happened – what they have allowed to happen. Then they placed a call to the real police, who arrived within minutes.

Ogborn's fiancé, Walker Nix, was found guilty of sexual assault and jailed for five years. Donna Summers received probation for false imprisonment. The original call was traced to a supermarket in Florida, where a police investigation led to the arrest of David Stewart, a thirty-seven-year-old guard working for a private prison company. Stewart was charged with impersonating a police officer and soliciting sodomy. By targeting businesses with strict procedural codes, Stewart had allegedly been able to find people predisposed to obeying authority and unused to handling novel situations independently.

There had been a spate of hoax calls at other fast-food restaurants: seventy similar cases were reported across thirty-two US states between 1995 and 2004. A caller to a McDonald's in Georgia convinced the restaurant's fifty-five-year-old janitor to perform a cavity search on a nineteen-year-old cashier. In North Dakota, a manager at a local Burger King was induced to strip-search a seventeen-year-old woman who worked under him. In Phoenix, a female customer at a Taco Bell was picked out and strip-searched by the manager. On the basis of these previous attacks, Ogborn sued her employers for failing to protect her adequately. McDonald's legal team argued that the hoax was a scam to make money, in which the teenager was a willing participant. The court rejected that story, and Louise Ogborn received $6 million (about £3.5 million) as a settlement.

David Stewart was acquitted of all charges. There wasn't evidence to prove that he was the person who had made the call. Since Ogborn's case hit the headlines, the hoax calls appear to have stopped.

MIND OVER MATTER

When physician and astronomer Franz Mesmer chanced upon hypnotism in the mid-eighteenth century, he furnished it with an abundance of extraneous detail befitting a showman – both theoretical and physical. Mesmer believed that the universe was saturated in a vital energy, the ebb and flow of which gave life to things, and whose tumultuous currents gave rise to all illness. He and he alone, he proclaimed, held the ability to manipulate these currents, a process by which a patient would become "Mesmerized" and cured.

To entrance a patient, the great Mesmer would stare into the person's eyes for hours on end, sitting so close that their knees bumped against each other. Simultaneously, he would rub their hands and press upon their chests. In the dim light, spectral music was played on a glass harmonica. Later, he devised a large vase run through with iron rods that could pipe this "animal magnetism" to multiple patients at once.

Hypnotism is still suffering from an association with such mystics and showmen but, to his credit, Mesmer was able to show that the body could be profoundly influenced by a mind open to suggestion. How profoundly? In 1845, physician James

Esdaile found himself in Calcutta as an assistant surgeon with the East India Tea Company, attending to the painfully swollen testicles of a convict from the nearby jail. Hoping to drain the accumulated fluids, he lanced one side of the scrotum, leaving the patient in agony.* Determined to find a less terrible solution before the follow-up surgery to the other side of the man's scrotum, Esdaile decided to mesmerize the patient, inventing his own ritual of passing hands over the man for hours while he lay quietly in a darkened room. The mesmeric surgery was a success, and soon patients were arriving from far and wide to submit to Esdaile's "painless" surgery. Despite claiming to have performed over three hundred operations using mesmerism, the practice petered out with the advent of chemical anaesthetics.

What is physically happening to a hypnotized brain that produces such altered states of consciousness, in which a person might feel no pain, or eat onions as if they were apples, or see colours where there are none? It has to do not with how we sense the world, but how we interpret those sensations. Writing in the *New York Times*, Sandra Blakeslee describes the neural process of being handed a rose:

> Photons bouncing off a flower first reach the eye,
> where they are turned into a pattern that is sent to
> the primary visual cortex. There, the rough shape of

---

* An outbreak of filariasis, a disease caused by tiny parasitic worms, meant that elephantiasis was a common complaint in Calcutta at the time. It is not known if they were suffering from the "occult form" of the infection (it does exist).

the flower is recognized. The pattern is next sent to a higher – in terms of function – region, where color is recognized, and then to a higher region, where the flower's identity is encoded along with other knowledge about the particular bloom... Bundles of nerve cells dedicated to each sense carry sensory information. The surprise is the amount of traffic the other way, from top to bottom, called feedback. There are 10 times as many nerve fibers carrying information down as there are carrying it up.

This downward flow of information, the rationalization of your sensory input, is strong enough to overrule the original input. Normally, this process allows us to make sense of incomplete or conflicting information, but it can also be exploited. Part of the trade of stage magic is to convince a person that a set of rules exist – that a steel ring is solid, that a box has no hidden trapdoors – so that our rational mind will turn a perfectly possible illusion into a seemingly impossible one. Likewise, hypnotism can exploit this mind-over-matter process to fool us into ignoring sensory inputs.

Clinical neuroscientist Dr Amir Raz neatly illustrated this overruling process in an experiment based on the Stroop effect. Named after John Ridley Stroop, this effect describes the delay caused when a person is confronted with conflicting pieces of information. In the classic version, cards with words written in coloured ink are flashed at a test subject, who is asked to name the colour of the ink as quickly as possible. That might sound easy, but when the cards show colour words (e.g. the word "red" written in blue ink), naming the right colour becomes a lot harder. Readers stutter, stumble and make mistakes, because the recognition of the word (a higher brain activity) overrides the sensory input of the visible colour.*

Raz hypnotized a small group of sixteen people – half of them identified as being resistant to the practice, half of them as highly suggestible – telling them that the words that would appear shortly were meaningless scribbles. Those sensitive to hypnosis performed much better in the Stroop test, ignoring the meaning of the word and concentrating on its colour: MRI scans of these test subjects showed that the region of the brain devoted to reading words was less active compared to the other group. Believing was seeing.

How could a man armed with nothing more than a glass harmonica and some metaphysical dressings exact the same sort of damage that the *burundanga* gangs of Colombia and CIA spooks could only attain with hallucinogenic cocktails? Perhaps the answer lies in the workings of the human mind. Rather than being a locked box that needs to be broken into – physically or chemically – it's an open game, a system that is constantly bombarded with information from outside, and it can be thrown off track with the tiniest of manipulations. Such innate suggestibility is vital to us as we try to understand the world, but it also puts chinks in our psychic armour. As Stanley Milgram demonstrated all those years ago, compliance is the greatest drug of all.

---

* Naturally, this only works if you can understand the words, making Stroop tests conducted in a language foreign to the subject a handy way of rooting out spies.

4
REMOTE / CONTROL

The individual is defenseless against direct
manipulation of the brain.

José Delgado, *Physical Control of the Mind* (1971)

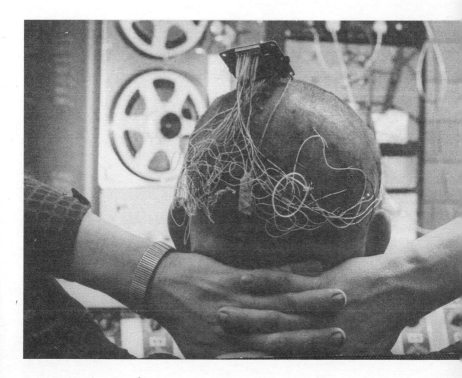

IF THERE EVER HAD BEEN A MODEL for the all-American boy, Charles Whitman would have been a contender for it. Born in 1941, his father was a successful plumbing contractor who applied the same ambitious drive to raising his family as he did to running his business. From an early age, young Charles was pushed to excel. By the time he entered school he could play the piano; an intelligence test taken at the age of six placed him in the top 0.1 percent of the country. He became an Eagle Scout at twelve, the youngest person in the world to do so, and served as an altar boy to his parish, Sacred Heart. Popular in high school, he played baseball and managed the football team. His gun-loving father introduced him to shooting, and by sixteen Charles was a skilled marksman.

After graduating, Charles Whitman enlisted with the US Marine Corps. He won a military scholarship to attend the University of Texas in the state capital, Austin. While still a student, he married. During those years, he once hauled the carcass of a deer back to his dormitory after a hunting trip – an act that was against the state's game-hunting rules, but which he shrugged off to the base chaplain as simply a "teenage prank". As part of his disciplining, the Marine Corps pulled him from his coursework and sent him to its training installation at Parris Island, South Carolina.

He did not let this disappointment interrupt his studies for long. When he returned to the university as a civilian, he held down a series of part-time jobs, collecting bills,

working at a bank, and surveying for the highways department, in order to support himself. He also volunteered as
a scoutmaster. He appeared busy and determined. By all
reports, Whitman was God-fearing, well spoken, handsome
and neatly dressed, "a truly outstanding person" according
to one of his supervisors. So it was a shock to many when,
on 1 August 1966, Whitman took a hunting rifle to the top
of a twenty-nine-storey tower on the university campus and
began shooting the pedestrians on the street below, killing
fourteen and wounding thirty-two. The killing only stopped
when he was gunned down by police.

The mystery of what had sent this odd but promising young
man on a killing spree unravelled during the subsequent investigation. Despite appearances, Whitman had been fighting
increasingly desperate battles over the terrain of his own mind.
He was plagued by fits of anger and, as he got older, these
attacks became more and more difficult to contain. During his
active military service he got into a fight, threatening to "kick
the teeth out" of a fellow marine over a gambling debt. For the
possession of a personal firearm on base, Whitman was court-
martialled, given a month-long stay in the brig, and demoted
several ranks to a private.

After enrolling at the university, he had scribbled copious
reminders to himself to remember to smile, to hold his temper,
to stop cursing, to remain calm. "Control your anger", he told
himself, "don't let it prove you the fool". He would embark
on passionate but cheerful discussions about the nature of
God with fellow students, but his fury always threatened to
leak through his contrived demeanour. One day, Whitman
lost his humour and physically threw another student out of

the classroom. Several times he had struck out at his wife. Mortified by these lapses in self-control, he had resolved to try even harder to suppress his temper. He started a diary, which he filled with long entries, having noticed that writing down his thoughts helped to clear his head. He composed heartfelt tributes to his wife, describing her as "the best thing I have in life". It was in vain.

Whitman was now plagued by headaches, which the huge quantities of Excedrin he swallowed did little to diminish. Bouts of insomnia kept him awake for days at a time, and he gained weight from frequent overeating. His violent urges grew stronger. In late July 1966 Whitman sought out the help of a staff psychiatrist at the university health centre. During an hour-long session, Dr Maurice Heatley noted that Whitman's outward civility seemed to be stretched taut over a massive well of hostile intent. At turns seething and despondent, Whitman spoke of his upbringing under a domineering and demanding father. He described his efforts to control his temper, and allegedly confessed that he felt like taking a hunting rifle to the top of the university tower to shoot people. The doctor asked Whitman to return the following week. By then, Whitman and sixteen others would be dead.

In addition to the fourteen people killed by gunfire on the university campus, two further victims were to be discovered. Before going to the tower, Whitman had attacked his mother in her penthouse, crushing her skull and stabbing her in the chest with a large hunting knife. He cleaned up the scene, laid the body in bed, and placed an apologetic note nearby. Whitman then returned home, slipped into bed next to his sleeping wife, and stabbed her with his knife, killing her instantly. What sets

Whitman's rampage apart from numerous other shootings through the decades are the details that surfaced soon afterwards.

The remorseful Whitman left a note confessing, "I cannot rationally pinpoint any specific reason for doing this." He also asked that any money remaining in his estate be donated to a mental health charity. "Maybe research can prevent further tragedies of this type," he wrote. Whitman's own suicide note, which investigators found on the tower's roof, read:

> I don't really understand myself these days. I am supposed to be an average reasonable and intelligent young man. However, lately (I cannot recall when it started) I have been a victim of many unusual and irrational thoughts. These thoughts constantly recur and it requires a tremendous mental effort to concentrate on useful and progressive tasks. I have been fighting my mental turmoil alone, and seemingly to no avail. After my death I wish that an autopsy would be performed on me to see if there is any visible physical disorder.

As it turns out, Whitman's suspicions about a physical disorder were correct. An autopsy later revealed a tumour the size of a golf ball lodged in the centre of his brain. It was a glioblastoma, a particularly aggressive and fast-growing class of brain tumour that is nearly always fatal. The growth had sprung up from Whitman's thalamus, and was pressing against his amygdala.

Could this mass of cells be what sent Whitman's brain out of control?

ALL IN THE BRAINS

The amygdala is a tiny, almond-shaped bundle of neurons nestled deep inside the brain.* Messages sent out from the amygdala communicate our most primal motivations; the primordial senses of fear, joy, arousal and anger all rise from this area.

These can grow so urgent and powerful that they overcome the neocortex, which governs our so-called executive functions – our cognitive control. For instance, when the right amygdala is artificially excited in a person, via electrodes, the stimulation can make the individual grow increasingly fearful or irate, with these emotions swelling until they can no longer be held back. As the dam of emotions bursts, the person usually directs an attack towards some external stimulus. He or she will make some effort to rationalize the behaviour – something must be blamed for the feelings coursing through the mind – but the person won't stop it. And even after the electrode is turned off, the emotions continue, often for a remarkably long time. Disturbances in the amygdala have also been linked to hypergraphia (compulsive writing), as well as strong feelings of religiosity.

The hypothalamus is a small region of the brain found in vertebrates. Located just above the brain stem, it plays a vital role in mediating cycles of sleep and appetite, amongst more complex body regulations. Lesions on some parts of the hypothalamus can cause hyperphagia (excessive hunger and intake of food). These symptoms of disruption to the amygdala and hypothalamus match the profile of Charles Whitman that was

---

* The Haitian *bokors* we met in chapter 1 might refer to the amygdala as something akin to the *gwo-bon anj*, the basic animating principle that drives us to action.

compiled after the University of Texas shootings, but the coroner's office also detailed far greater problems.

The tumour had also reached the neocortex's frontal lobe. The frontal lobe is associated with our capacity for self-control, forward planning and the mediation of impulses. This is an important talent to possess if we have any hope of arresting our actions until we've had a chance to consider their consequences. When someone shoves past you in the street, it's this ability for reflection that quells your urge to retaliate with your fist.

Damage to the frontal lobe can have dramatic consequences. Perhaps the most famous case of frontal lobe injury concerns Phineas Gage, a construction foreman who was working on the US railroads in the middle of the nineteenth century. One of Gage's duties was to blow apart large boulders that were lying in the way of the tracks; his work involved gunpowder. He would drill a hole into the offending rock, pour gunpowder into the hole, lay a fuse on top of the gunpowder, and then press the fuse down into the gunpowder with a large iron rod. On 13 September 1848, this chain of events went awry, and the charge ignited as Gage hammered his tamping rod into the hole. The blast propelled the rod upwards through Gage's cheek and tore clear through his skull, landing some distance away. Amazingly, he survived. Afterwards, friends reported changes in his personality. He acquired a habit of swearing prodigiously; he grew obstinate, impatient, selfish and unrestrained. He allegedly took to gambling and sleeping with prostitutes, vices in which he had not previously indulged. Once, Gage had been a studious and calculating operative; now he was child-like in attitude and unfit to work. It appears that Gage accidentally lobotomized himself, shearing off a good portion of the frontal lobe in the left hemisphere of his brain.

Despite the extraordinary case history of Charles Whitman, the case is in no way conclusive when it comes to indicting the cancer cells for his rampage. This was a man under immense social and academic pressure; biology was not the only factor at work. Failings in his military career and poor performance at school, exacerbated by the expectations of a domineering father and the disintegration of his family, had stressed Whitman to breaking point. There are numerous shooting sprees we can point to that were prompted by nothing other than mental stress, but as the coroner duly noted, the impact of Whitman's tumour cannot be ruled out, and he surely displayed a range of symptoms that make it a compelling theory.

Whitman's tumour was an accident of nature, Gage's lobotomy was an accident of technology. But what if someone set out to physically tinker with a brain in order to alter its owner's personality? Could you create your own army of willing zombies by suppressing the higher functions of the brain while simultaneously taking control of amygdala- and hypothalamus-triggered instincts? We don't have to guess at an answer, because others have already tried to do it.

CUT TO THE ASYLUM

The gruesome spectre of the dancing corpse of the hanged George Foster was haunting the public's minds when the German neurologist Eduard Hitzig hunkered down in the cramped kitchen of his Berlin apartment with some dogs: it was here that Hitzig, a man said to be in possession of "incorrigible conceit and vanity complicated by Prussianism", aimed to raise galvanic stimulation out of the jurisdiction of showmanship and into the dominion of worthy science. His results, published in 1870 in an essay entitled "On the Electrical Excitability of the Cerebrum", signalled a landmark in the emerging field of neurobiology.

Hitzig, assisted by the anatomist Gustav Fritsch, had been experimenting on dogs for some time. The two researchers had sliced through dogs' scalps to expose the brain and then stimulated areas of the grey matter with a thin metal needle, provoking muscle contractions. The deans at the University of Berlin were horrified. The tests were banned from the university campuses, and the two researchers were forced to decamp to Hitzig's home. There, they positioned the dogs on a small dresser table to receive more electric shocks.

As they carried out their research, Hitzig and Fritsch made particular note of the parts of the brain that they were stimulating and the result. The predominant question facing neurology at the time was that of "localization": did certain parts of the brain have particular functions, or was the bulk of the brain unspecialized, a lump of undifferentiated neurons that tackled all tasks equally? Other scientists were already hard at work trying to pinpoint the role of parts of the brain. Experiments carried out in rabbits and pigeons by the French physiologist Jean Pierre Flourens demonstrated that removing the brain stem caused death, while the wrinkled part at the back of the brain, known as the cerebellum, played a part in balance and motor co-ordination. The cerebral hemispheres were "the seat of volition and sensation", according to Flourens' finding. By shaving thin slices from the hemispheres, he had observed a gradual decline in rational faculties, leading him to believe that the functions of this lobe were distributed evenly throughout its whole mass.

It was time for Hitzig and Fritsch to dig deeper. With a dog sprawled across his dresser, Hitzig steadily worked his probe from a point at the top of the brain down to the side, noting how stimulating the left part of the brain provoked movements in the right side of the dog's body. In this way, the duo were able to chart the "motor strip", a patch in the middle of the brain responsible for voluntary movement in the body.

This strip appears in roughly the same place in humans, and it reveals that the neurons are not organized in the brain to follow the organization of cells in the rest of the body. If you were to draw the figure of a person on the brain so that each limb was adjacent to the neurons that controlled it, you would end up with a figure draped across the head, its legs dug into the top of the brain and its body hanging downwards. In a curious twist, its face would appear disconnected from the head, appearing somewhere below the supine torso.

The amount of brain devoted to each part of the body is markedly different, and depends on the level of fine motor control needed. To give you a better sense of the brain power involved, let's imagine redesigning the human body so that each limb is proportional in size to the area of the cortex that serves it – what scientists call a "motor homunculus". This wonderfully creepy little troll would stand atop spindly legs; its skinny torso would be corked by a gigantic head – or, rather, a small head with a gigantic face. The homunculus's eyes are massive, like dinner plates, and his huge, thick-lipped mouth opens to reveal a colossal tongue. Even more impressive are the hands, so large that the homunculus could wrap its entire body in their embrace.*

Hitzig and Fritsch had made headway, but other neurobiologists sought a more ambitious target – the very seat of the soul. Stimulating the motor strip might cause the leg to shoot out, but as Giovanni Aldini and Andrew Ure had shown a century earlier, so too would applying an electrode to the nerves running along the leg. In other words, this was a mechanical consequence, much like a bullet flying out of a rifle at the pull of a trigger. What interested the next generation of scientists was the trigger itself: the impetus to carry out an action. There must be a difference between the physical impulse that leads to a muscular movement and the thought required to generate that impulse. Indeed, both accident and experiment had proved that damage to the brain could leave a person or animal alive and perceptive to the world but in a state of heavy lassitude, lacking motivation. In such cases, it seemed, the *corps cadavre* remained alive but the *gwo-bon anj* was lost.

At the time of Hitzig and Fritsch's experiments, the best attempt thus far to reach the seat of the soul in the brain seemed to have been made by the German physiologist Friedrich Goltz. Goltz had demonstrated that removing the neocortex of dogs resulted in increased aggression, from which he concluded that the neocortex acted to inhibit this behaviour. Yet Goltz himself eschewed the idea that brain function was localized. At

---

* You can see one example of a motor homunculus at the Natural History Museum, London. He typically stands alongside his cousin, the sensory homunculus, who boasts similar proportions but is noticeably more, let's say, well endowed.

an international meeting of doctors held in London in 1881, he showed that a dog with portions of its brain removed was still functionally a dog. Ultimately Goltz would find himself on the losing side of the argument, with the localization of brain function firmly entrenched as neuroscience dogma at the end of the nineteenth century.

News of Hitzig's brain map travelled rapidly around the medical community, as did word of Goltz's experiments with aggressive dogs. At the Préfargier asylum, situated on picturesque Lake Neuchatel, in Switzerland, psychiatrist Gottlieb Burckhardt paid special notice. Burckhardt was director of the stately institution, whose magnificent views and well-tended gardens sought to establish a calming atmosphere for the residents. Burckhardt believed that psychological disturbances came about when parts of the brain misfired; by severing the connections between the source and the destination of these errant signals he hoped to suppress "bad" behaviour – a treatment for the symptoms, but not a cure. He started to wonder whether humans might have their aggression literally cut out, using Hitzig's brain map as a guide. He decided to call the innovative surgery a "leucotomy", from the Greek for "white cut", a reference to the white nervous tissue of the brain.

At Burckhardt's instruction, a small three-room surgery was built at the asylum where he could perform his new-fangled technique. Unfortunately, the operations were a disaster. Of the six patients that went under Burckhardt's scalpel in 1888, one died within a week, another committed suicide some time after, and two showed no change in psychiatric profile. The lucky two who experienced a decrease in symptoms also experienced heavy side effects such as a decrease in speech ability.

The mother of one of these patients told doctors that he had become "quieter, better behaved, and more manageable" – an ominous prognosis that was a trademark of psychosurgery for the next century. When Burckhardt finally shared his results publicly, other physicians were critical of both his methods and his theory, and he abandoned the practice altogether.

Burckhardt's work and name were largely forgotten until almost fifty years later, in 1935, when the idea caught on again, this time in a sterile operating theatre of the Hospital Santa Marta in Lisbon. Overseeing the procedure was the celebrated Portuguese neurologist António Egas Moniz, trialling a new technique he called "prefrontal leucotomy" in a clear nod to Burckhardt's earlier work. Moniz, assisted by Almeida Lima, drilled a hole into each side of the patient's head, and then used a syringe to inject alcohol directly into the brain, destroying the connection between the prefrontal cortex and the thalamus. It was clumsy work. The team later improved upon the technique using a device that resembled a builder's sealant gun and which they dubbed the leucotome. The instrument featured a thin nozzle with a loop of stiff wire that could be used to slice through brain tissue.

After operating on fifty patients, Moniz felt that he had achieved a significant level of success, especially in treating mood disorders. He published the results in several medical journals and presented his findings to a group of international colleagues in London, drawing widespread press coverage.

At the time, the treatment of mental disorders lagged woefully behind advances in other areas of medicine, and it seemed as though Moniz had made the long-awaited breakthrough that had been out of Burckhardt's reach. Not everyone was enamoured with Moniz's techniques, however. In 1949, a disgruntled patient shot him four times with a pistol; one of the bullets lodged in Moniz's spine and he was confined to a wheelchair for the rest of his life. The injury prevented Moniz from travelling to Stockholm that December to accept a Nobel Prize for his work in psychosurgery.

Leucotomies carried with them a great measure of risk – even Moniz agreed with that. The surgeons operated blindly, slicing at parts of the brain carried out mostly on a hunch. Sometimes the surgery had no effect, at other times it reduced the patient to a vegetative state. It was, quite literally, a hit-or-miss practice. Psychiatrists had precious few tools to work with, though, and were open to considering dramatic, haphazard interventions for patients plagued by persistent and crippling illnesses such as depression and schizophrenia. Desperate times called for desperate measures. In a lengthy monograph from 1891, Burckhardt defended himself against his critics, writing:

> Doctors are different by nature. One kind adheres to the old principle: first, do no harm (primum non nocere); the other one says: it is better to do something than do nothing (melius anceps remedium quam nullum). I certainly belong to the second category…

Leucotomy was a delicate surgery, restricted to well-equipped hospitals – those best situated to do something rather than nothing. And if things had stayed that way, the procedure might have remained fairly low on the radar of medical history. That was not to be.

EXCISING THE SOUL

As Moniz presented his findings at the 1935 Second International Neurological Congress in London, an American neurologist named Walter Freeman sat in the audience, spellbound. Immediately on returning to his lab at the George Washington University (GWU) Hospital, in Washington, DC, Freeman began preparations to replicate and refine the operation. The following year, assisted by his colleague James Watts, he carried out America's first leucotomy. The patient was Alice Hood Hammatt, a sixty-three-year-old housewife from Kansas who suffered from crippling anxiety. After the surgery, she was "able to sleep without medication and live without a nurse's care" for the first time in years. Her husband said that the next years were "the happiest of his wife's life".

When the men presented their achievement at the annual meeting of the Southern Medical Association in Baltimore, Maryland, *Time* magazine reported that doctors could "cut the ability to worry out of the brain". The *New York Times* proclaimed it "surgery of the soul". The scientists attending the meeting were more reserved in their judgement. Expressing the unease that many doctors felt towards the procedure, eminent psychiatrist Adolf Meyer commented prophetically: "I have hesitations at the thought of a great many of us having our distractibility or our worries removed. To call attention to these operations may start such an epidemic."

Over several years, the GWU group gradually improved and standardized their version of leucotomy, renaming it the Freeman-Watts procedure – what was to become known as the lobotomy. Even with the refinements, things remained complicated and somewhat haphazard. Instead of a leucotome,

Freeman and Watts used a dull knife with a rounded tip (think of a butter knife), which was inserted through slits drilled in the temples and used to cut away at the brain. Over the course of 624 lobotomies, Freeman and Watts observed good results in 44 percent of their patients, with poor results showing up in 28 percent (not counting the particularly poor result of death, which occurred 3 percent of the time). As might be expected, given the history of psychosurgery, what Freeman and Watts counted as an improvement was open to debate. Nurses reported that patients became infantile after the operation, and had to learn again how to speak and go to the bathroom. Others were said to have been reduced to docile automatons who paced meaninglessly in circles, or spent hours staring at the wall. These dangers were well recognized – when William Seabrook met the zombies working the plantation in 1928, he compared their blank expressions to those of lobotomized dogs he had seen once in a laboratory. Across the board, however, an estimated one-third of patients who underwent a lobotomy improved enough to be discharged from psychiatric institutions.

The psychiatric community remained uneasy about the use of lobotomies, given the range of results. Over half of patients experienced side effects such as partial paralysis, incontinence, obesity and convulsions, not to mention psychological disturbances such as sexual dysfunction, a loss of social awareness, reduced vocabulary, word deafness and apathy. In 1941 Freeman famously carried out a lobotomy on twenty-three-year-old Rosemary Kennedy, sister of the future US president, to treat her violent mood swings; the botched operation left her in a mentally incapacitated state for the rest of her life.

As other treatments emerged and the relative costs of the lobotomy mounted, the procedure was restricted to the worst affected patients. Freeman himself claimed he wouldn't consider patients for lobotomies "unless they are faced with disability or suicide". Still, colleagues noted that Freeman displayed a somewhat cavalier behaviour when it came to his work. Just before Alice Hood Hammatt's lobotomy, she had changed her mind about the operation when she discovered it involved shaving her head. Freeman told her he would save as much of her hair as he could. He later crowed that when Hammatt awoke from the surgery and saw that he had sheared her cherished locks, "she no longer cared". The *Washington Post* offered an even more disturbing version of events, reporting that Hammatt was forcibly anaesthetized, crying out before her lobotomy, "Who is that man? What does he want here? What's he going to do to me? Tell him to go away. Oh, I don't want to see him."

Cavalier or not, Freeman passionately believed that lobotomies would serve to alleviate the strain on the country's mental asylums, which at that time housed over half a million patients. However, so long as lobotomies required a trained surgeon working in an operating theatre, the procedure was only available to a small number of patients. Few clinics could afford the necessary facilities, or could gain regular access to hospitals with such facilities. This motivated Freeman to look for another route into the brain, one that didn't require cutting through the skull.

An Italian psychiatrist named Amarro Fiamberti had carried out a lobotomy by entering the brain through the eye sockets, where the bone of the skull is thinnest. Pondering this at home, Freeman picked up a grapefruit from a bowl of fruit and began stabbing at the skin with an ice pick. If he were to sink a blade through the eye socket, a lobotomy could be performed without conventional surgery. Officially, the technique was known as the trans-orbital lobotomy; the press would call it the "ice pick lobotomy". After putting the patient under anaesthesia, the thin tip of an ice pick was inserted above the eye, resting against the thin bone of the skull.* A hammer blow then drove the pick into the skull, where it was jerked from side to side, and up and down, to slash at the tissue of the brain. Complications were reduced and mortality rates fell by half. To Freeman, it was the advance he needed in order to bring lobotomy to the masses.

The trans-orbital lobotomy decreased side effects and could be carried out in as little as ten minutes by psychiatrists with little surgical training. This meant that the procedure could be performed quickly and cheaply in asylums across the country. Around this time, Freeman fortuitously uncovered a notable trend among his patients. The people who responded best to lobotomies, he said, were those who had been institutional- ized for less than six months. By Freeman's estimate, over two-thirds of these recent patients went on to live productive lives outside of hospital following the operation. By contrast, long-term residents of psychiatric institutions showed little benefit – the chance of a good recovery was less than one in ten after spending seven years in care. In Freeman's view, this meant that for a lobotomy to be successful, it had to be carried out as soon as possible after a patient was admitted. Even if a

---

* Freeman later used a specialized pick, which he confusingly named a leucotome; when one of these snapped off inside a patient's skull, he fabricated a stronger version – the "orbitoclast".

patient's symptoms were relatively minor – neuroses rather than psychoses – Freeman deemed it sensible to carry out a lobotomy. Why wait for things to get worse? Whereas once he had claimed only to carry out lobotomies on those in danger of "disability or suicide", Freeman now insisted to his critics that "it is safer to operate than to wait".

From the outset, Freeman aggressively marketed these new services directly to patients, including hiring out billboards. In a single fortnight, he carried out more than two hundred trans-orbital lobotomies on patients at West Virginia mental hospitals. Other doctors voiced their concern, insisting that lobotomy should be considered as the last resort in treatment. James Watts, the man who had been at Freeman's side in bringing leucotomy to the US, was so disgusted with his colleague's zeal that he ended their partnership. But Freeman was obsessed. Within eighteen months he had conducted over five hundred lobotomies in West Virginia. Many of these patients were then discharged from the state's institutions, which meant huge savings for the government. The authorities were more than happy to let Freeman continue selling his wares.

Like a tent revivalist or a patent medicine man, Freeman took to the road, preaching the gospel of lobotomy – the miraculous cure-all for mental ills. On the day of a surgery, he would parade a lobotomized animal around town and swing a rattle over his head to draw a crowd. Patients would be lined up in a row, with Freeman racing from one bed to the next to show how swiftly he could perform the operation. He even carried out lobotomies on both sides of the brain at once, holding a pick in each hand. As his geographical range expanded, so too did the range of illnesses that could be safely treated with lobotomy. Almost

any complaint could be answered with the ice pick – depression, anxiety, alcoholism, headaches, restlessness, criminal behaviour.*
Notoriously, Freeman lobotomized a twelve-year-old boy named Howard Dully at the behest of his new stepmother, who felt that the boy's daydreaming and general misbehaviour warranted the treatment. Dully later wrote in his autobiography that he spent his life feeling "like a freak".

The circus came to an end in 1967 when Freeman was contacted by Helen Mortensen, one of the earliest recipients of his trans-orbital lobotomy. She had undergone a second procedure after her symptoms returned. Now sixty-three, she had regressed once again and asked him to perform another lobotomy. With an orbitoclast still embedded in her skull, Freeman stood back to pose for a photograph, accidentally knocking the pick with his elbow. The jolt cut open a major blood vessel in Mortensen's brain and she later died of the injury. Freeman's licence to practise medicine was revoked. Over the course of his career, he had carried out around 3,500 lobotomies, 70 percent of which were conducted with the ice pick.

It might be that the medical authorities had grown tired of Freeman's antics and saw the opportunity to put an end to them. Freeman's credibility had been in decline for over a decade, in

---

* Freeman was a man of radical ideas. In 1930, he was interviewed by *Time* magazine about his work in the autopsy room of St Elizabeths Hospital, also in Washington. It was Freeman's firm belief that different personality types predisposed themselves to particular diseases – a kind of twentieth-century Humourism. "Schizoid types", for example, were pale, sharp-featured loners who often suffered from tuberculosis. By contrast, "cycloid types" were round-faced and jovial, and prone to heart disease.

part because other treatments were beginning to compete with his surgery. Although initially used in large doses, which were criticized as being a "chemical lobotomy", chlorpromazine was effective and treatment could be scaled down or stopped entirely, a marked improvement to psychosurgery.

THE JOY MACHINE

Walter Freeman tried to silence the unruly parts of the brain by cutting them out with a knife. But for other scientists, neurosurgery held the promise of much more sophisticated engineering work.

Although Eduard Hitzig had failed to isolate the exact seat of the soul in the human brain, his map was still of some use. Following in his footsteps were surveyors ready to stake claim to the new geographies of the brain, and to make them their own. These scientists embraced a mechanistic philosophy of the brain, in which the mind is nothing more than an elaborate clockwork, a billion cogs turning together in perfect harmony – or, in patients suffering from mental illness, less than perfectly. If a gear ran too fast, or ground against another, then what was needed was a precise tool, something far less crude than a butter knife.

One psychiatrist who set about building that toolkit was Dr Robert Galbraith Heath, who founded the Department of Psychiatry and Neurology at Tulane University in New Orleans. Heath's first great discovery in mental health was an antibody in the blood of schizophrenics, which he called taraxein. Using healthy human volunteers, Heath claimed that the protein was capable of producing temporary schizophrenia-like symptoms in them. He was never able to isolate the antibody, nor was anyone else able to replicate his findings. Nevertheless, it encouraged Heath's belief that disorders of the mind were no different to diseases of the body, in that both must stem from some type of biological injury or imbalance.

Heath's chief contribution to the treatment of mental illness was his development of electrotherapy, the technique of surgically implanting electrodes in the brain. Through this he hoped to map the limbic system, the area deep in the brain responsible for generating feelings and emotions. Using dentistry drills, Heath and his assistants cut tiny holes in the skulls of their patients, then gently pushed long needles into their brains. But unlike with Freeman's lobotomy, the goal was not to obliterate the neurons there but to record them. Using these implanted needles, Heath and his team measured the patient's brainwaves. They found that patients in the midst of a psychotic rage or catatonic stupor showed patterns of brain activity markedly different from those of healthy people. "By implanting electrodes and taking recordings from these deep-lying areas, we were able to localize the brain's pleasure and pain systems", Heath reported. "We'd interview a patient about pleasant subjects and see the pleasure system firing. If we had a patient who flew into a rage attack, as many psychotics did, we'd find the 'punishment' system firing." Perhaps most importantly, Heath discovered that the pleasure and pain pathways operated in opposition – when one was activated, the other was suppressed: a binary system. All one had to do to suppress violent episodes in schizophrenic patients was to use electrodes to stimulate the pleasure zone in the brain. Heath commented that: "The primary symptom of schizophrenia isn't hallucinations or delusions. It's a defect in the pleasure response. Schizophrenics have a predominance of painful emotions. They function in an almost continuous state of fear or rage, fight or flight, because they don't have the pleasure to neutralize it."

This may sound like a plot element from a sci-fi thriller, but Heath's techniques were therapeutically successful. Over the next twenty-five years, Heath and his colleagues operated on more than sixty patients, inserting up to 125 electrodes in a single individual. Generating a tiny current in these needles activated the surrounding neurons. When the pleasure zone was stimulated, patients suddenly giggled like children, even though they could not identify what it was that amused them. Some patients were given "self-stimulators", which allowed them to elicit doses of neural pleasure whenever they felt the need – basically, they were given a "joy box". Whereas rats given similar devices self-dosed until they passed out from bliss, humans displayed more restraint (although one man managed to clock up fifteen hundred doses during a three-hour session). With these electrodes in place, Heath was able to create a permanent implant that discharged mediating pulses of electrical activity straight into the pleasure zone, dampening the uncontrollable anger that some of his patients expressed.

The first recipient of this neural pacemaker was a mildly retarded man whose fits of anger made him "the most violent patient in the state", known for repeatedly trying to slash himself as well as his caregivers. Electrodes that sent out calming pulses every five minutes were implanted in the man's cerebellum, which Heath had discovered was an efficient place to access the emotional pathways of the brain. These electrodes were connected to an external battery about the size of a cigarette packet.* After the operation, the man abruptly ceased his violent attacks and was well enough to be allowed home. However, one day he suddenly reverted, seriously wounding a next-door neighbour and attempting to murder his parents. Subdued by police and returned to hospital, X-rays revealed that the wiring of his neural pacemaker had come loose. Once repaired, the man's violent behaviour disappeared once again.

It was by no means a cure-all, but Heath estimated that about half of his patients benefitted from the neural pacemakers. Naturally, the doctor's ability to put the mood of his patients on remote control aroused interest from certain branches of government. At some point, a representative from the CIA allegedly approached Heath and asked if he could stimulate the pain zone in the human brain in the same way he could manipulate the pleasure zone. A device that could remotely and invisibly inflict pain on a person would be quite useful for interrogation – even for behaviour modification, mind control. Heath threw the agent out, calling the request abhorrent. "If I wanted to be a spy, I'd be a spy", he boomed, in an account of the conversation to the *New York Times*. "I wanted to be a doctor and practise medicine".

Other researchers appeared to be less discriminating about the application of their efforts, which is handy if you want to direct someone by remote control.

---

* Later models had batteries small enough to be implanted along with the pacemaker, providing for a less awkward set-up.

CONSTRUCTING A PSYCHO-CIVILIZED SOCIETY

In 1963, a young scientist named José Manuel Rodriguez Delgado stepped into a bullring in Cordoba, Spain. Opposite him was a particularly aggressive young bull. When Delgado waved the matador's iconic red cape, the animal thundered towards him, its hooves throwing up great clods of dirt and sawdust. Delgado didn't flinch. At the last possible instant, he pointed a small box at the animal and pressed one of its buttons. The bull stopped its charge and stood placidly in front of him. Delgado was spared! To many watching, it seemed an extraordinary feat, a miracle. But to Delgado it was simply the debut of his newest invention.

Lodged inside the bull's brain was a "stimoceiver", a type of radio-linked neural pacemaker that Delgado had designed in his laboratory in the physiology department at Yale University. Using this device to activate electrical brain stimulation (EBS), he could make the animal dance and bellow on command. He claimed that EBS allowed him to suppress the bull's aggression as well as its movements.

Many of the spectators probably wondered who this impudent matador was, a man who could defeat a bull without touching it. Delgado had worked with many patients suffering from neurological conditions, such as epilepsy, which did not respond to any conventional treatment. During these investigations he had found, like Heath, that he was able to manipulate emotions in his test subjects, most notably their sexual desire.

It happened first with a thirty-six-year-old woman who was being treated for epilepsy. While she was undergoing a post-operative therapy session, each implanted electrode was excited in turn, as a therapist conducted an interview. As her right temporal lobe was stimulated, she reported pleasant tingling sensations down one side of her body. Repeated pulses increased her pleasurable sensations; the patient became chatty and flirtatious. At one point, her desire was so provoked that she asked to marry the therapist. A second patient, "an attractive, co-operative, and intelligent" woman of thirty, experienced similar effects when an electrode in her right temporal lobe was fired. She took fondly to a therapist attached to the project (whom she had not met before), kissing his hands and expressing how grateful she was for what he had done. In sessions held before and after this application of electrical brain stimulation, both women were reserved and poised, displaying none of the overly friendly behaviour seen during the EBS trials.

It seemed clear that the EBS had elicited the pleasurable sensations, but Delgado was careful to distinguish between the sensations and the patients' attraction towards the researchers who were with them. He wondered if this attraction might be explained by some underlying "female strivings". Could his neural pacemakers actually make people fall in love? Or did they simply reduce a person's inhibitions?

As a point of comparison, he considered the case of A.F., an eleven-year-old boy who had undergone a similar brain stimulation procedure to treat his severe epileptic fits. As the machine cycled through a prearranged sequence of pulses, the boy cried out: "Hey! You can keep me here longer when you give me these; I like these." His left temporal lobe had been stimulated. Later in the session, the boy began to question his sexual identity, telling the therapist, "I was thinkin' if I was a boy or a girl – which one I'd like to be." Then, when the electrode in his temporal lobe fired, he blurted out, "You're doin' it now. I'd like to be a girl."

Without a compelling medical application, prospects for a button-activated joy box wired directly into a person's head fell away after Delgado's experiments. Nerve stimulation did persist as a treatment for a range of conditions such as Parkinson's, epilepsy, depression and pain, however. So it was that in 1998, a North Carolina-based doctor named Stuart Meloy once again accidentally stumbled upon the powerful erogenous effects of EBS. This time, the patient was a woman suffering from chronic back pain. The patient had opted for a procedure known as sacral nerve stimulation, where an electrode is implanted into the spine. During the operation, as Meloy tested the placement of the electrode, the woman released a ragged gasp. "You'll have to teach my husband how to do that", the woman exclaimed to Meloy. Intrigued by the effect, he mentioned it to some other doctors. One suggested that the stimulation might be used to treat women suffering from sexual dysfunction. He tested an electrical implant, which he dubbed the Orgasmatron, in eleven women and two impotent men. The device, which is adjusted using a remote control, is currently undergoing human trials. Meloy might have a hard time persuading everyone that a low sexual appetite – which is about as specific as a diagnosis of "female sexual dysfunction" – justifies invasive surgery.

Back when Delgado pioneered EBS in the 1960s, he wanted to improve more than libido; he was flirting with the idea that it could improve other, non-bedroom relations too. One experiment saw Delgado presiding over a population of gibbons, whose brains had been radio-chipped, as they roamed across a small island in the Bermudas. Using his control box, he inhibited aggression in certain individuals in the group, restructuring the group's social order. He even gave the remote to a low-ranking female gibbon to see if she would use it to her advantage. The female learned how to suppress the alpha male of the group and would turn him on or off according to her whim.

Delgado believed that such social manipulation would be beneficial to human society. He was a veteran of the Spanish Civil War, during which he had served as a medic with the Republicans. Having seen the worst excesses of man, he was preoccupied with the idea of reining in these excesses. He suggested that a radio-linked implant could detect the change in brainwaves that occurs when a person experiences a psychotic event. In Delgado's scheme, this information would be broadcast to authorities, who would send a signal back to the implant via an FM radio transmitter in order to incapacitate the person before he or she could do any harm. In his 1969 tome *Physical Control of the Mind: Toward a Psychocivilized Society*, Delgado cited neurotechnology as holding promise for conquering the mind and producing "a less cruel, happier, and better man". It was not a popular book. The prospect of using brain implants to control human behaviour evoked the worst excesses of fascism rather than protections against it, but Delgado insisted that: "this danger is quite improbable and is outweighed by the expected clinical and scientific benefits". The scandal grew larger when two of his co-researchers suggested that the techniques might be used to quell the race riots plaguing America's inner city areas. Public attitudes towards EBS soured, and Delgado's funding dried up.

Delgado's legacy lives on, however. Programmable animals haven't entirely disappeared; DARPA, last heard from in chapter 1, where it was funding research into suspended animation, also maintains the Hybrid Insect Microelectromechanical Systems (HI-MEMS) programme. Its goal is to create insects that can

be guided by remote control, acting as living, breathing surveil-
lance bugs. Cornell University researchers have been able to
take control of hawk moths in flight. On the other side of the
US, scientists at the University of California, Berkeley, posted
a video on YouTube that showed a fist-sized giant flower beetle
take flight and veer about a room following the commands tapped
into a laptop. Going one better, in 2007 Su Xuecheng and a
team of fellow scientists at the Robot Engineering Technology
Research Centre, at Shandong University of Science in China,
created remote-controlled pigeons. The neuroengineers were
able to steer the flying birds left, right, up and down, thanks to
microchips embedded in their brains.

But is this mastery over the mind or over the body? Are the
pigeons compelled to turn against their will, or is a desire to fly
in that particular direction sparked in their unwitting brains?
As it happens, Delgado himself carried out several experiments
to test whether he was controlling volition or movement. He
observed that animals whose movements he controlled showed
no signs of distress – as they did when they were physically
restrained – and they sometimes held an induced pose even
after the stimulation ceased. From this he concluded that the
animal did not perceive the movements as forced.

If you, say, implanted a stimoceiver in someone, could he or
she resist the commands sent to it? That would depend on the
part of the brain affected and the strength of the impulses sent.
One of Delgado's patients was asked to resist impulses relayed
to his motor cortex that were meant to make his hand curl into a
fist. After several failed attempts, he sighed, "I guess, doctor, that
your electricity is stronger than my will." Delgado noted that,
for the most part, electrodes implanted into the motor cortex

produced clumsy, unnatural movements; the ability to control a complex motion such as walking would, he believed, require the precise, co-ordinated action of dozens of electrodes – something that neuroengineers today are working to make a reality.

It is far easier to stimulate parts of the brain to inspire emotions – which a person might likely act upon, all the time perceiving the behaviour to be voluntary. Delgado, for instance, repeatedly induced a form of anxiety in one patient, triggering him to begin searching his immediate surroundings. Unlike the man who had electrodes implanted in his motor cortex, this patient always reported that his searching was a spontaneous action, generated in his own mind. He would offer excuses such as "I heard a noise" to explain why he'd set out, and gave no indication that he felt the behaviour had been provoked by external forces. Yet, it is impossible to tell whether Delgado planted into the patient's mind the suggestion for the search, or if the patient was compelled to act out of a general restlessness and then made excuses for why he was acting in that way. So the seat of the soul remained as hidden to Delgado as it had to Flourens.

Although complex behaviours had been induced, the technique remained unreliable. Even today, scientists have observed a level of control limited to simple commands. This suggests that the circuitry of the human brain varies from person to person, making wholesale mind control via stimoceivers unfeasible on a wide scale. But while taking control of complex behaviours is currently beyond the ability of us humans, it is not beyond the ability of other species.

5
THE GHOULISH NANNY

Ill can he rule the great that cannot reach
the small.

Edmund Spenser, *The Faerie Queene* (1596)

WHEN ANTÓNIO MONIZ INJECTED ETHANOL
into the brain of his patient back in 1935, he could hardly have
known that his new-fangled lobotomizing procedure had already
been perfected by somebody else. That earlier pioneer was the
tiny insect *Ampulex compressa*, known more familiarly as the
emerald cockroach wasp. An elegant, glittering metallic blue-
green beauty of the entomological pageant, the wasp displays its
more brutish nature when it comes time to give rise to its next
generation. Wasps are a nuisance, as the old ditty goes, and bees
are worse, but if your foremost complaint about wasps is their
ability to ruin a good picnic, spare a thought for the cockroach
wasp's victims.

When creepy-crawlies have nightmares, they must be about
wasps. Members of this insidious order of insects turn spiders into
cocoon-spinning slaves, and burst out of the innards of maggots.
The smallest insect known to science, the wasp *Dicopomorpha
echmepterygis*, is smaller than an amoeba and lives inside the eggs
of the barklouse. If no fellow insects are about on which to prey,
plants will do for the oak gall wasp, which manages to induce
tumours on the leaves of oak inside of which its larvae grow,
protected from predators. Or rather, they would be safe there,
were it not for the *other* wasps that thrust steely probes into the
gall and lay eggs that attack the larvae. Wasps are everywhere
and in everything, sparing none.

The specialty of the emerald cockroach wasp is reproductive
parasitism. When the female wasp is ready to lay an egg, she

searches high and (mostly) low for a suitable partner to help raise her offspring. The ideal partner is not another wasp, however, but a cockroach. She spies one and flies close. Compared to the lumbering, well-armoured roach, the wasp is small, and she cannot subdue it with strength alone. She stalks her prey with care, waiting for the perfect moment to leap onto the roach's carapace. A brief struggle ensues, but the wasp usually manages a clumsy strike, thrusting her stinger into her mate's abdomen. The venom paralyses the roach, but the effects only last for a moment. With little time to spare, the wasp positions her stinger over the roach's head and delivers a second, far more precise sting, driving her slender stiletto probe directly into the brain. She then seeks out the part of the nervous system responsible for the roach's escape reflex and douses it in venom. When the local paralysis induced by the first sting wears off, the cockroach remains stupefied. Experiments have found that the cockroach is able to right itself when turned over, indicating that its motor skills are intact and it is not paralysed. Instead, the insect appears to lose its *will* to walk (when shocked with small amounts of electricity, or placed in a water-filled container, it makes little or no attempt to flee). Now that the docile cockroach is under the wasp's control, the wasp seizes it by its antennae, using them almost like reins, and, in the words of one researcher, leads the roach "like a dog on a leash" to a prepared burrow. There, the wasp lays its egg on the belly of the roach and seals the tomb from the outside. The last step is not so much to keep the cockroach in as to keep unwanted predators out; the wasp has the utmost faith in the fealty of its zombified roach nanny.

The cockroach remains in a semi-comatose state until the wasp egg hatches. At this point, the wasp larva chews its way into the roach's body, devouring the internal organs in a selective order to guarantee its living victim survives for as long as possible. Several days later, the adult wasp erupts from the corpse of the roach, ready to begin the cycle yet again.

As Vera Cosgrove, aka Mum, noted admirably in Peter Jackson's *Dead Alive*, aka *Braindead*, "No one will ever love you like your mother!"

LIVING LARDER

For a long time, parasites were largely overlooked by scientists. Perhaps that's because few schoolchildren grow up wanting to devote themselves to mapping the life cycle of the nematode worm. Grander, gaudier animals, such as the big cats of the African savannah and the multicoloured birds of the Costa Rican rainforests, are far more appealing to most of us. It is the dandy, the dangerous and (in recent times) the endangered that attract the human brain's attention. But parasites need watching too, especially if zombiism is your goal.

Far from being degenerate, dependent creatures, parasites are among the most sophisticated and treacherous organisms on our planet – and we encounter them *every day*. Mammals, birds, reptiles, amphibians, fish, insects, plants, nematodes, fungi, algae, bacteria: all are prey to infection. Almost every free-living species has an associated parasite, and most have quite a few. In fact, parasitism is the dominant form of life on Earth. It is not the exception, but the rule.

Parasites finally received a small nod of recognition in 1980, when Peter W. Price shook up the scientific community with his revolutionary *Evolutionary Biology of Parasites*. In it, he suggested that the difference between a predator and a parasite was merely one of selectivity. A fox is equally happy to dine on a rabbit, a hen or a cat; in contrast, a monarch caterpillar can only feed on the leaves of the milkweed plant. In this respect, Price said, the caterpillar is no different from a tapeworm dependent on the gut of a specific host. Precipitously, the world was packed with parasites – they were everywhere to be seen. But Price's argument was about more than semantics. He also predicted that sets of organisms locked in host–parasite complexes would be highly adapted to each other, having been engaged in an evolutionary "arms race" as generations of hosts tried to defend themselves against the parasite and the attackers countered these defences.

If you've ever felt that the world hates you and your zombie-making ambitions, spare a thought for parasites. They must exist in (or on) a living being that detests their presence in the most fundamental of ways. The immune system of a host reacts instinctively to any unwelcome settlers with an arsenal of physical, chemical and behavioural weapons designed to eject or kill them. But like any settler forging a trail into a new frontier, parasites do everything within their power to shape the hostile environment around them and build a comfortable home. They need food, warmth, safety, transport, a mate and a chance to reproduce, just like any other creature. For a non-parasite, securing these things can be as simple as weaving a nest or digging a network of tunnels. Parasites, however, must often weave their nests from their host's living tissue in order to take advantage of the bounty of resources available in that body. And with a great deal of ingenuity, parasites learn how to avoid or overcome the elaborate defence systems of their host, making almost any environment habitable.

It's one thing to resist the attacks of a host environment, but parasites are able to go better, not just defending against their host but reprogramming it, at the host's expense, to better suit the parasite's own needs. It's within such adaptations that scientists have since discovered nature's most incredible examples of zombiism and mind control. Host manipulation takes on myriad forms, from subtly influencing existing chemical processes to full-on enslavement and destruction. The manipulation doesn't have to be complicated. When threadworms (a nematode of the genus *enterobius*) travel to the end of the human digestive tract to lay their eggs, they secrete irritating chemicals that ensure the eggs are transferred onto the hands of their itchy host, and from there the parasite can spread to other humans. Guinea worms (*Dracunculus*) are quite capable of moving through the human body undetected, yet when a mother reproduces, her offspring incite a painful burning sensation and blisters*. The easiest way to soothe the affected skin is to bathe it in cool water. When this happens, the young worms break out of the host and swim away to complete the next stage of their life cycle. Host manipulation doesn't have to be chemical, either. Mites of the genus *Antennophorus* have perfected a trick of stroking the mouthparts of their ant hosts. For an ant, this is like having a pair of fingers pushed down the throat, and they regurgitate their food, which is then eaten by the mite. And these are relatively benign parasitic tactics, as the emerald cockroach wasp and its offspring have already demonstrated.

---

* *Dracunculus* as in "little dragon", the same root from which "Dracula" arises.

Most if not all parents want to give their offspring the best possible start in life, and parents that happen to belong to parasitic species are no exception. Just because some animals do not take a direct role in parenting does not mean that they abandon their children to the indifferent world. Quite the opposite, in fact – these parents do well to ensure their young are born into an environment that is as safe and nourishing as possible. This is why so many insects lay their eggs on rotting meat and vegetable matter, so that their larvae have a plentiful supply of food on hatching. A rotting carcass might not seem like an appropriate nursery for our own offspring, but to the larvae of many insects it's Disneyland and McDonald's rolled into one. This strategy is not without its risks, however: food can dry out, be washed away, scavenged or taken over by micro-organisms that cause rot and mould. In addition, larvae born on a food patch out in the open will have to compete with the young of other animals that choose to lay their eggs there too, and the whole cosmopolitan population of eggs and larvae will be at constant risk from roving predators.

Depositing eggs on fresh food promises a solution to many of these problems. So you can see why many parasites prefer to lay their eggs on a living host – what could be a fresher source of food? A species of solitary wasp, known as beewolves (*Philanthus triangulum*), take the fresh-food approach, laying their eggs on live bees. When the larvae hatch, they bore into their still-living host and digest it from the inside. Naturally, bees aren't too keen on this arrangement and attempt to remove the eggs should they find such a dubious gift deposited on them. To stop the larvae from being dislodged, the adult beewolf, much like the emerald wasp, paralyses the bee with the venom in its sting, then places the bee in a burrow – part larder, part nursery. Conscious that all pantries are prone to mould (even living ones), the beewolves douse the zombified bees with antibiotic *Streptomyces* bacteria, cultured in a set of specialized glands in the wasp's abdomen. The paralysed bees remain fresh and well-preserved right up to the moment when the beewolf's eggs hatch.

Beewolves aren't the only parasites to have come up with a strategy for disarming the dangers of choosing a living host. The Carolina sphinx moth (*Manduca sexta*) spends its childhood as a big fat green caterpillar commonly known as the tobacco hornworm, which, as the name suggests, makes a regular meal of tobacco leaves. As a hornworm contentedly munches on its nicotine-laced lunch, the tiny female *Cotesia congregata* wasp dives down and uses its sting-like ovipositor to pepper the larva's soft skin with wounds, each one now home to a single wasp egg. As it implants the eggs, the wasp also injects a symbiotic virus that knocks the caterpillar's immune system out of action. When the eggs hatch, they live inside the hornworm, feeding off its internal juices. After an incubation period, the wasp larvae are ready to spin their own cocoons, inside which they can mature into adults. This poses a problem, however, because they must leave the safety of their caterpillar host, who would be more than happy to switch from a diet of tobacco leaves to one of wasp cocoons. But the larvae are able to adjust the hornworm's levels of the neurotransmitter octopamine. This places the caterpillar into a coma-like state; it will never reach its own cocoon stage and go on to become an adult moth, but instead will be stuck as a caterpillar, growing fatter and fatter as the digestive system is stalled. How they maintain this control after leaving the hornworm is still a bit of a mystery. One theory is that some larvae stay behind (by accident or design) and continue to wield a subduing spell over the insect, allowing their brethren to mature safely.

PARASITE PUPPETEERS

Incapacitating a host and devouring it is one thing – easy zombie work. It's another altogether to coerce a host into behaviours that should never occur normally. How do you get your zombie to heed your beck and call? A lesson might be taken from parasitic wasps of the genus *Glyptapanteles*, which is found in the Americas.

Glyptapanteles are famous among scientists for using zombie nannies to guard their young. The familiar tale begins like that of the ill-fated tobacco hornworm: a *Glyptapanteles* wasp attacks a caterpillar of the geometrid moth (*Thyrinteina leucocerae*), using its sting-like ovipositor to embed upwards of eighty eggs into the caterpillar's soft skin. Inside, the eggs hatch and the wasp larvae grow, feeding on the host, which ignores the strange crawling sensations inside itself and continues feeding as normal. Once mature, the larvae burst out of the caterpillar's skin and weave cocoons nearby so they can metamorphose into adults.

But the caterpillar's role in raising the young wasps is not yet over. The cocoons make a welcome snack for passing predators such as beetles, and during this vulnerable time the ex-host is recruited to stand watch over the helpless larvae. Should any beetles approach, the caterpillar thrashes about fitfully, spitting noxious fluids to drive away the threat. Experiments carried out by researchers from the University of Amsterdam and the University of Viçosa in Brazil found that pupae watched over by a zombified caterpillar suffered half the losses of unguarded cocoons. The team, led by Amir Grosman, are still unsure exactly how the larvae cast this spell over their former host, but they suggest that one or two larvae are left behind inside the caterpillar to control it from within. So diligent a watchman is the caterpillar that it doesn't even stop to eat during its long guard-shift. Shortly after the adult wasps hatch out of their cocoons, the caterpillar perishes.

Slave nannies crop up with surprising regularity throughout the natural world. When dreaming up the sources of zombie plagues, Hollywood's writers have often turned to the animal kingdom: dogs, crows, rats, woodpeckers, worms and even a "Sumatran rat-monkey" have been enlisted as infectious actors. So far, there hasn't been a blockbuster with a barnacle villain, but perhaps it's time for the movie industry's scientific advisers to correct that. The *Sacculina* genus of barnacles is a particularly malicious parasite of crabs that takes control of both the mind and body of a host in order to raise its young. After squeezing through a gap in the female crab's exoskeleton, the barnacle travels to the crab's underside and takes up residence where the crab's eggs would normally be incubated. From this point on, the crab becomes a surrogate mother for the parasite offspring, treating the *Sacculina* eggs as if they were its own. The crab loses the ability to perform normal functions, such as moulting, producing eggs or regrowing limbs; the parasite dictates that the crab's energy be funnelled directly to the young barnacles. Once the barnacle offspring are ready to be released into the sea, the crab climbs a rock and releases the larvae in clouds, stirring the water with its claws to aid their dispersal.

It's not easy for a *Sacculina* parasite to tell the difference between male and female crabs, and often they will infect male crabs, which lack the maternal instincts of the females. This is a problem that the barnacle has no problem overcoming. If it infects a male crab, it simply interrupts the normal hormone balance to "castrate" the male and turn it into a mother. The barnacle interrupts the male crab's hormonal system, provoking physical changes such as a widened, flattened abdomen that can better host the eggs, as well as behavioural changes – the male will even venture out to sea to brood, though it is no longer able to produce sperm. Having lost its ability to reproduce, the crab becomes little more than a fleshy extension of the parasite. It's a perfect example of what Richard Dawkins calls the "extended phenotype", a behaviour that maximizes the survival of the genes responsible for it, even if those genes are not part of the animal itself.

Octopamine – the neurotransmitter used to take over the tobacco hornworm – seems to play a key role in making zombie nannies. Researchers at Ben-Gurion University in Israel found that by injecting a mimic of the neurotransmitter into the brain of the roach, they could "revive" cockroaches that had been zombified by the emerald cockroach wasp, suggesting that the wasp's venom acts by blocking this chemical. When the initial paralysis wears off, the cockroaches stop to groom themselves thoroughly rather than instantly running away from their unwelcome master. Octopamine appears again and again in insect (and crustacean) biology – for example, helping with faster relaxation of leg muscles in locusts, sparking light production in fireflies, fuelling social behaviour in bees and controlling aggression in fruit flies. It would not be surprising if it were a key ingredient in changing crab sexuality. Octopamine is also related to noradrenaline, which is involved in the fight-or-flight response.

Little is known about how octopamine affects human behaviour, although it does have physical effects. Chronic use of MAOI-type antidepressants causes octopamine levels to rise in the sympathetic neurons, which can cause dizzy spells; when mixed with other stimulants, octopamine can also result in high blood pressure (the chemical is often found in weight-loss supplements, owing to the belief that octopamine encourages the burning of fat).

Whether or not injecting octopamine inhibitors into the brain might turn humans into zombies remains to be tested. The trick will be in finding a big enough wasp. But there are some very big parasites already conquering the earth beneath our feet.

THINKING CAPS

Fungi represent one of the five taxonomic kingdoms with an estimated 1.5 million species worldwide. They are perhaps the most enigmatic of organisms, even more so than wasps: they are able to thrive in almost any environment; share traits with plants and animals, but are distinct from both life forms; and can display as many as sixteen different sexes. Far from the image of a docile, sessile button-cap mushroom, or even a thrill-generating, mind-altering psilocybin variety, some fungi are carnivorous, even predatory; certain species use nets and loops to capture tiny nematode worms. Of course, it stands to reason, then, that many fungi are parasitic, and a great many show expertise in altering the behaviour of their hosts – and not just by putting them in a groovy mood. They too put their victims to work.

Ophiocordyceps unilateralis is one such parasitic fungus, whose species name hints at the single-minded bullying it employs.

Its host of choice is ants. A tiny spore of the fungus enters the ant through the insect's breathing spiracles and it soon begins to consume the non-essential soft tissues of the ant as a meal. When the fungus is ready to spore, it pushes bundles of tiny threads called mycelia into the ant brain, and through these puppet strings it takes control of its host, compelling it to climb a nearby plant or tree and, once high enough, fasten itself there with its powerful jaws. Only after the ant is secured in place will the fungus kill its host. A final growth spurt forces bundles of fungal threads to burst out of the ant like the wool of an overstuffed plush toy. From the ant's neck, tall fruiting bodies sprout, their swollen heads heavy with spores. By willing the ant to climb a plant, the fungus enjoys a much higher vantage point from which to shower infectious spores onto its next potential hosts, down below.

Ants are all too aware – if you define "awareness" in evolutionary terms – of the danger posed by *Ophiocordyceps unilateralis*. Any ant showing signs of infection is quickly carried by other members of the colony as far from the nest as possible and expelled, and for good reason: an outbreak of *Ophiocordyceps* fungus can easily wipe out an entire colony – a clear lesson for any human victims of a zombie outbreak.

There are more than four hundred known species of *Ophiocordyceps*, every one of which has a corresponding host that it parasitizes, including dragonflies, cockroaches, aphids, beetles, stick insects, butterflies, bees and wasps. It has competition, too. Another fungus, *Entomophthora muscae*, infects and kills houseflies in an almost identical manner. After an infectious spore lands on a host, the fungus threads itself through the body of the fly, absorbing its nutrients. Infected flies are easily recognizable by their grossly distended abdomens, which become swollen with this unwelcome cargo. In humans, this kind of infection might lead to a fever, where the body's temperature rises in an effort to cook out the invading pathogen. Insects can't control their body temperature in the same way, but the fly has been able to develop a similar trick. It moves to a higher spot where it will receive more light, and thus more heat. Unfortunately, this is just what the parasite wants: a high vantage point from which to spread its spores. As with *Ophiocordyceps*, the fungus then kills its host and sprouts spore-loaded tendrils from the corpse, from whence erupts a "conidial shower" (as scientists refer to it, charmingly). If these conidia do not land on a susceptible host, the fungus has one last trick up its sleeve: the large spores develop into smaller "secondary conidia", which are more likely to be picked up by the wind and spread further afield.

You might think that these manipulations of a host are nothing more than crude lobotomizations (or in the case of *Sacculina*, the appropriation of an existing behaviour to serve predatory needs). Could your zombies be trained to do something more than that – perhaps even learn an entirely new skill? One of the more intriguing of parasitic nanny modifications arises in the relationship between the larvae of the *Hymenoepimecis argyraphaga* species of wasp recently discovered in Costa Rica and its host, *Leucauge agryra*, an orb-weaving spider. Like the emerald wasp, the adult *Hymenoepimecis* wasp stings the host to temporarily paralyse it, during which time it lays a single egg on the spider's abdomen. When the spider recovers it continues life as normal, unaware that the egg has hatched and a small wasp larva is clinging to its belly, feeding on body fluids that drip from small holes it has chewed in the spider's abdomen. This is pretty conventional parasitism, as it goes.

Once the larva is ready to move from its juvenile existence into adulthood, something far more unusual happens: the spider begins to spin a web unlike any that it has created before. Instead of the familiar wide net, the web is a robust plinth supported by thick anchor lines. This new design isn't a sudden burst of arachnid creativity; the architecture is engineered by the parasitic wasp larva. To become an adult, the larva needs to create a cocoon in which to metamorphose, and it directs the spider to create a suitable platform – something solid and robust that won't be damaged by falling rain. When the spider finally completes the new web, the parasite kills its host and ceremoniously eats it. It drops the spider's empty corpse to the ground and moves to its specially crafted platform, where it waits until nightfall before making a cocoon.

Scientists are trying to uncover how the larva achieves this feat. It seems that the parasite can provoke the spider into repeating a subroutine that is normally only one small part of regular orb construction, and suppress all other normal web-building behaviour. What is most troubling is that the *Hymenoepimecis* larva is an external parasite, which means it achieves this mind control with some kind of poison, injected far from the host's brain. Experiments have shown that if the wasp larva is removed during the special web's construction, the spider will continue to weave the platform for several days before it reverts to normal spinning, indicating that the venom is both fast-acting and long-lasting.

Still, the exact nature of the poison has left scientists in knots to explain it.

THE KNOT OF THE PROBLEM

Once upon a time, the people of Phrygia were without a king. When they consulted the oracle at Telmissus, in what is now modern-day Turkey, they were told to appoint as their new king the man next arriving by cart into the capital. That unexpected coronation fell to a poor peasant named Gordius. In thanks to the powers that allowed such an arbitrary change in his family's fortune, Gordius's son Midas dedicated the ox-cart to Zeus and fixed it in front of the palace with an elaborate knot. It was such a fiendishly complicated knot that no one could undo it, and it was said that whoever managed that feat would go on to rule all of Asia. It took the sword of Alexander the Great to fulfil the oracle's prophecy and release the cart from its Gordian knot.

But Gordius's legacy was not limited to string alone.

Gordian worms are parasitic creatures that measure just a few millimetres in diameter but can grow to over a metre in length, giving them the appearance of long strands of flaxen hair – and the nickname "horsehair worm". During mating, they contort themselves into intricate tangles akin to the mythical knot. Armed with tools much more complex than a bronze sword, scientists have been able to untangle the curious mysteries of the worm's lifestyle.

The adult worm leads a short, unremarkable life, wallowing in murky freshwater (it's particularly fond of man-made water troughs), in something like a hangover nap from its wild adolescence. The juvenile, in contrast, is a particularly nasty specimen that infects, digests and enslaves its host before finally killing it.

When the tiny gordian worm larvae hatch from their water-dwelling parent, they cannot survive long without finding a terrestrial host in which to mature to adulthood. Using an armature of spines on the proboscis, they bore into the flesh of whatever animal happens to be passing by close enough. If they happen to burrow into a fish, snail or crustacean, they never make it further in the life cycle; these waterbound hosts are a dead end. The lucky ones find an insect, such as a mosquito or dragonfly, in its waterborne larval stage. When the insect metamorphoses into a flying adult and leaves the pond, the hitchhiker worm catches a ride, encased inside its body by a protective cyst. On land, the gordian worm continues to develop, and to look for new hosts to prey upon. Should its first insect host be eaten by another invertebrate – say, a spider, beetle or mantid – the worm embarks on the next stage of its gruesome life.

For one species of the worm, *Spinochordodes tellinii*, that stage usually takes place in an insect of the order Orthoptera, which includes grasshoppers and crickets. Once its original mosquito host has been devoured by a cricket, the protective cyst encircling the worm larva breaks open and the larva penetrates the its new host's gut, where it secures itself. The worm immediately begins to secrete enzymes that effectively digest the animal from the inside out. Over several weeks, all the time feeding on the unfortunate cricket, the larva develops into an adult. As the adult's body reaches lengths several times that of the cricket's, it packs itself into tight coils that fill the host's increasingly hollow body cavities.

To complete its life cycle, *S. tellinii* must return to the water. This poses something of a Gordian problem, because the parasite is lodged inside a terrestrial host, and crickets are not known for their love of swimming. If the worm were to burrow out of the cricket, it would find itself stranded on dry land. To overcome this seemingly intractable issue, *S. tellinii* opts for its own "Alexandrian" solution: it takes control of the cricket's mind and convinces it to commit "suicide". Against its nature, the zombified cricket ambles about until it comes across a body of water, then falls or leaps into it.

To find out if the parasite was truly steering the grasshopper, a bit like Delgado steering his bull with his stimoceiver, French biologist Frédéric Thomas of the Centre for the Study of Polymorphism of Microorganisms in Montpellier took up sentry duty at the edge of a private swimming pool in southern France and spent the balmy summer evenings catching crickets. Some were collected in the nearby forest, while others were taken from within one hundred metres of the pool. Every night four test subjects (two found in the woods, two from nearer the pool) were placed at the water's edge to see if they would leap in. After fifteen minutes, all the crickets were recaptured and preserved for dissection. Only 15 percent of the crickets captured in the forest were infected with gordian worms, compared to 95 percent of those found near the pool. Given the size of the gordian worm compared to the cricket, it's worth noting here that four of the crickets were found to have two worms living inside them, while one unhappy cricket was infected by four of them. Of the infected crickets, almost half made a leap into the water, whereas only 13 percent of uninfected insects did the same. The results suggest that the infected crickets were attracted to the water, both at a distance and close up.

Yet, Thomas was not convinced, so he devised a second experiment. A Y-shaped maze of clear plastic was constructed in the lab, offering crickets two destinations: a dry trough or one filled with water. Small fans wafted air through the corridor so that the crickets could discern a humid breeze coming off the water-filled container. The crickets had thirty minutes to find their way into one of the troughs before the experiment was concluded. If they could track down a swimming pool from the woods, this should be easy. But there was no preference amongst the infected crickets for the water-filled trough, which was visited by infected and uninfected crickets alike. Only the infected crickets, however, felt the urge to throw themselves in once they'd arrived. Thomas concluded that the gordian worm sent the crickets wandering aimlessly, and that it was only by luck that the insects found themselves near a water source. Once there, the worm somehow convinced the cricket to throw itself in.

Little is known about how the worm achieves its control over cricket behaviour. Recently, a group of French scientists led by David Biron, the editorial advisor of the aptly named *Open Parasitology Journal*, reported that when the worm was inhabiting the host cricket, it produced molecules that affected the proteins of the cricket's brain, namely those involved in neurotransmitter function and motor control in response to gravity. The parasite also mimicked cricket proteins relating to the development of new nervous system tissue and circadian rhythms (as well as several that weren't identifiable).

Like the manuscript of some great composer, we are slowly identifying some of the notes used, but it's still not clear what it sounds like when you put them all together. For now, the worm's Gordian knot remains unpicked.

YOUR BODY IS A BATTLEGROUND

Given the sheer abundance of parasites, we shouldn't be shocked that Thomas discovered four gordian worms tangled around each other in that unfortunate cricket – we've got a whole lot more parasites lurking inside our own bodies. Parasites may make their beds in living flesh, but that doesn't guarantee a world free from competition. In nature, a resource worth having is a resource worth fighting for, and hosts are no exception. So what happens when two parasites decide to plant their eggs in the same host?

Roderick Fisher of Cambridge University decided to find out by establishing a parasitic death match to take place within the confines of a tobacco moth. The fighters in this 1961 rumble were two ichneumon wasp species that reproduce by laying their eggs inside the larvae of the tobacco moth. Using a syringe, Fisher carefully injected one egg from each wasp into the caterpillars, then left nature to take its course. In all cases where the parasites were allowed to mature without intervention, only one emerged victorious from its host. To find out what happened, still more moth larvae were infected and dissected at key intervals. What Fisher found was a brutal and often bloody fight to the death between the two wasps.

On hatching, both swam furiously around the inside of the caterpillar, duelling each other. The battle ended with one wasp larva biting the other with its sharp mandibles, leaving puncture wounds that darkened and leaked the larva's own bodily fluids into those of its host. The wounded parasite, no longer hidden from the caterpillar's immune system, was quickly entombed in a cyst and died. Things became more thrilling (to an ento-mologist) when Fisher stacked the odds in favour of one wasp over the other. By implanting the eggs at an interval, with one several days older than the other, he found that the older parasite always won when they were locked in combat. But even more interestingly, given a sufficient head start, the older parasite won without any physical combat at all. Given a fifty-hour head start, the older wasp parasite was able to suppress the growth of the competing wasp. As soon as the newcomer hatched from its egg, it shrank and died, victim of some unknown physiological suppression, either by an excretion from the existing parasite, or a change in the host environment (i.e. the loss of key nutrients). The tobacco moth larvae can have the dubious peace of mind in knowing that if it has been parasitized by one ichneumon wasp, at least it won't be parasitized by another.

And as we'll see, this kind of superparasitism is important, because parasites are not simply focused on taking over their hosts; they want to take over the world.

6

ARMY OF BLOODSUCKERS

There is nothing so patient, in this world
or any other, as a virus searching for a host.

Mira Grant, *Countdown* (2011)

SO YOU WANT TO CREATE YOUR OWN ZOMBIE ARMY. Perhaps you've fashioned your own Delgado-style brain implant out of some kebab skewers and a couple of AA batteries. Or maybe you've gained the ability to take control of your neighbours using nothing more than the sound of your own authoritative voice. Or you've discovered a furtive protein that can be secreted directly into your target's mind, forcing them to go on an unwanted night-time swim. Unfortunately, none of these stratagems lend themselves to mind control on a grand scale. Wrapping just one or two people in a dissociative fugue via phone requires constant, careful, personal manipulation. Even if you managed to persuade a large mass of people to install mind-control chips or proteins in their heads, the logistics of precisely controlling each of them would be maddening, if not impossible. Humans are simply so complicated that it takes most of our brain power to control our own body – any more than that is quite a feat.

What is needed then to launch your zombie army is a *distributed network*: a plague of the shambling dead that can sustain itself, even organize itself, without your direct involvement.

Could such a thing be achieved without human intellect overseeing it? As it turns out, nature has evolved its own zombie armies, which take advantage of the strengths of distributed networks, and they're far more complex and advanced than anything humans have managed. As we saw in the last chapter, parasites can infest all manner of creature, and the tiniest can be the most nefarious. That is especially true when one looks through the microscope, to see protozoa that hide themselves in cloaks made from our own cells, or viruses that write themselves into our DNA.

SEALED WITH A KISS

If we were to roam the world as a troupe of mad scientists, searching for the ultimate zombie bug, where might we go first? Perhaps we'll set our plane down in South America, in Brazil or Chile, to meet the assassin bug. These small beetles live in burrows, nests and anywhere else that's not too far from a warm-blooded companion. They emerge at night to feed on their sleeping host, the bugs' soft, flat bodies swelling up with each blood meal. Those living with humans tend to feed on their host's face, a bedside manner that lends them another name – the kissing bug.

However, the kissing bug isn't the candidate for our mad scientist zombie plague; we're actually interested in something that lives inside these beetles. Just as the bugs feed on us, tiny little worm-like creatures called trypanosomes feed on them. The single-celled parasites in our kissing bugs are *Trypanosoma cruzi*, and *T. cruzi* infect humans as well as insects. Kissing bugs use your face not only as a dinner plate but also as a toilet bowl, and the faeces they leave behind contain microscopic spores of *T. cruzi*. When these spores are absent-mindedly rubbed into an eye or a small cut, infection occurs, leading to Chagas disease, a debilitating long-term condition that damages the heart and digestive system beyond repair. Beetles harbouring *T. cruzi* bite twice as often and defecate more frequently than beetles without, helping to spread the disease-causing microbe. In other words, *T. cruzi* doesn't simply take advantage of the insect's blood-sucking lifestyle to spread from one victim to the next; beetles infected by the tiny creatures are transformed into voracious eating machines.

Are the beetles truly zombified or are they simply eating more to compensate for the loss of nutrients siphoned off by the parasites inside them? To answer that we can turn to a related trypanosome, *Trypanosoma brucei*, which is responsible for African sleeping sickness, and uses the tsetse fly to spread from one victim to another. Scientists at the Institute of Development Research in Montpellier wanted to know more about how *T. brucei* affects the tsetse fly behaviour. Looking at the proteins found in the heads of infected flies, the team found that twenty-four different proteins had been switched on or off, including ones involved in control of the central nervous system. It was as if the scientists had discovered *T. brucei*'s toolkit lying open right inside the flies' brain. A microbe that knows how to turn bugs into ravenous feeding machines – we'll put that in our mad scientist bag.

BAD AIR IN THE DOLDRUMS

The next step on our round-the-world jaunt is a small tropical island that is little more than a speck of green in the endless blue of the Pacific. Here, a woman lies deep in the grip of malaria. The back room where she has been quarantined from her family is empty but for a mountain of blankets. Just visible under the heap is the woman's face, slick with fever. Her mouth hangs loosely in a slow pant; her eyes wide open but caught in a fog of delirium. Her blood is thick with animal-like, single-celled parasitic protozoans of the genus *Plasmodium*, which have been injected into her by a tiny female *Anopheles* mosquito, itself sick with the parasites. As the mosquito stabbed its syringe-like proboscis into the woman's skin to feed on her blood, it released dozens of microscopic sporozoites. The woman's immune system has been slow in responding to this threat: while feeding, the mosquito also injected a cocktail of drugs to prevent the wound from clotting, and these chemicals had silenced the nerve endings that would have alerted the body to this breach in its defences. By the time that the woman felt the familiar itch of a mosquito bite, the offender was long gone, and the sporozoites were rushing to her liver. Within thirty minutes, they had hidden themselves there and were reproducing asexually, each sporozoite dividing itself into two new cells so that the population was doubling, generation after generation, roughly every forty-eight hours.

A couple of weeks later, the sporozoites developed into a new form, merozoites, and burst out into the woman's blood. As they did, they wrapped themselves in cloaks sewn from the membranes of her own liver cells, protecting themselves from her immune system. These merozoites attacked her red blood cells, gorging on the haemoglobin molecules inside and reproducing again, occasionally bursting out of one cell and invading another. It is only now, during this assault, that the symptoms of these invaders finally show: fever, weakness, joint pain, anaemia, vomiting, convulsions – malaria. Even if the woman survives this bout of the illness, there is a good chance that the parasite will return after a period of incubation, sometimes as long as thirty years later.

But the infected zombie in this story is not the woman under the mountain of blankets, it is the *Anopheles* mosquito that carried the malaria to her. We can say for sure that the culprit was female, because only female mosquitoes take blood meals. Although all *Anopheles* feast on nectar, a snack of blood is required to build up a protein essential for egg production. We can also say that it was an elderly mosquito, because *Plasmodium* parasites take almost the entire lifespan of the mosquito to develop.

Plasmodium can't travel from one human to another by themselves, they need to hitch a ride on a passing insect. They use the mosquito as a "vector" to get from one person to another. In its youthful days, our female mosquito grabbed a meal of blood from someone already infected with malaria. This was not completely accidental: that person smelled wonderful to the young mosquito. The *Plasmodium* inside them, using a technique we don't yet understand, were able to alter the person's odour subtly, by changing the chemistry of his or her sweat, or perhaps the breath, to make it deeply attractive to passing mosquitoes.

The parasites, swallowed whole as the insect fed, bored through the mosquito's gut wall, where they enclosed themselves in cysts. Then something curious happened. Normally, the mosquito will make regular meals of blood and nectar in its efforts to lay as many eggs as possible during its two-week lifespan, but mosquitoes infected with *Plasmodium* lose their appetite. This is the parasites' doing: they're not yet ready to be passed on to another human, and they're not ready to risk ending their own life cycle. Every time the mosquito bites a person, it runs the risk of being swatted to a pulp. For the mosquito, that risk of death is a price worth paying to supply its offspring with nutrients. The on-board parasites disagree, so they suppress the mosquito's appetite until they have divided and matured into sporozoites, strong swimmers that can invade a human host's liver cells. Thousands of these infectious forms travel to the mosquito's salivary glands, where they sit like bullets in the chamber of a revolver, primed to be shot into a new human host. Only then will the parasites release the mosquito from fasting.

When the infected mosquito found our patient, something else strange happened. An uninfected mosquito takes its blood meals as quickly as possible and then beats a hasty retreat; the longer it remains attached to a human, the more chance it has of being crushed to death. But the mosquito that bit our patient adopted a leisurely manner, feeding for longer than usual. This too was *Plasmodium*'s doing. Once they've come this far, they have no vested interest in the mosquito's survival: this bite could be the only chance the parasite gets to find a human host. The parasites deploy some form of molecular deception that whispers reassurances to the mosquito, the exact nature of which biologists still do not understand. Somehow it convinces the mosquito to risk its own life to serve the parasites' needs by staying longer on the host. These tricks allow the *Plasmodium* to improve their chances to spread and prosper, and they are very successful at this: these single-celled protozoans are responsible for infecting 225 million new victims each year, 800,000 of whom will die from the resulting malaria.

With mosquitoes as a vector, a single human infected with *Plasmodium* can easily spread malaria to thirty others, each of whom will spread it to thirty more. In this way it's easy to see how a single bite can unleash an epidemic. It also explains how malaria continues to dodge the best vaccines that humans can make: the sheer number of people exposed to an infection means that it can find the gaps in a vaccination programme – those who missed their shot or didn't respond to it – and pass through them like water running through a sieve. In essence, *Plasmodium* have created their own inoculation programme, staffed by billions of dedicated volunteers in a distributed network that spans the globe.

Could a similar, specially evolved microbe turn other species into rampant bloodsuckers for you? The average human brain weighs around 1.3 kilograms (three pounds), more than a million times the size of that found in a kissing bug or a mosquito. Humans have something on the order of one hundred billion neurons whose branching, intertwined dendrites create over a thousand trillion synaptic connections. That's a lot of grey matter to control.

But don't presume that "advanced" animals with complex brains are somehow immune to parasitic meddling, or even that they are more difficult for a host to manipulate. Parasites can take control of creatures many times larger and more complex than themselves; it's what they evolved to do, after all. And the manipulation that they have perfected is at a level that not even humans, who like to believe they are the smartest, most rational beings on the planet, can escape.

INFECTED WITH RAGE

On 3 October 1849, the famed horror writer Edgar Allan Poe was found sprawled outside Ryan's Saloon in Baltimore, Maryland, incapacitated by delirium. A physician was called and Poe was transported to Washington College Hospital. The normally well-dressed author was clad in shabby, ill-fitting clothes; his shoes were worn through. He could not explain his state. Days earlier he had visited his family home in Richmond, Virginia, and proposed to his childhood sweetheart, before setting out to attend to some literary business in Philadelphia. Yet his luggage had never left the Swan Tavern in Richmond. And there was no record of any instructions left by Poe for its delivery elsewhere.

In the hospital ward, he was racked with tremors and hallucinations, fading in and out of consciousness. When told that a friend would soon arrive to visit him, Poe moaned that the best thing his friend could do for him would be to blow out his brains with a pistol. He became aggressive, and the hospital orderly was forced to tie him to his bed with heavy leather straps. Four days later, Poe was dead. He was, according to Dr J.F.C. Handel, Baltimore's commissioner of health, a victim of "congestion on the brain".

The circumstances surrounding Poe's death have inspired much speculation. Could he have died from an attack of cholera or syphilis? Perhaps he was a victim of "cooping", a nineteenth-century practice in which a man was abducted by a gang, drugged and then coerced into casting votes in multiple political districts – a conjecture that might explain Poe's shockingly altered appearance and his delirium. Or was his notorious alcoholism or drug use to blame? Despite his proclivity for these, Poe had been abstinent for a year, during which time he was healthy. He did not match the profile of a typical addict, deteriorating slowly under the burden of a vice. It was a mysterious death perfectly suited to the master of the macabre.

Years later, in 1996, a doctor at the University of Maryland Medical Center, in an effort to solve the mystery of Poe's death, decided to present this complex case at the hospital's weekly clinical pathology conference. When the anonymous history of symptoms and death were shared with the assembled physicians, cardiologist R. Michael Benitez put forward the theory that the patient had succumbed to rabies. The patterns of alternating delirium and lucidity, his aggression, his inability to swallow fluids (whether alcohol or water): all were hallmarks of late-stage infection with the virus. Patients admitted to hospital in the final throes of rabies typically do not last long; on average, they die within four days, often from an inflammation of the brain. Of course, Benitez didn't have to look far for inspiration for his theory – Poe's grave lies just one block from the medical centre. He soon wrote up his novel analysis, though there is no means of confirming it as no autopsy was conducted.

Rabies spreads itself using some very simple and brutal tactics, making it particularly successful and fearsome.* Typically, the parasite that causes it, *Lyssavirus*, is introduced to a new host through a bite: a single drop of saliva contains enough virions – individual virus particles – to produce an infection. At just 0.0002 millimetres, these bullet-shaped organisms are one-tenth as long as a *Plasmodium*. Once inside the body, they locate a peripheral nerve and begin to inch along it, using the nerve like a guide rope that leads the parasite all the way to the host's brain, where it begins to wreak its damage. This journey can take as many as several months, depending on how far from the head the bite occurred.

---

* The disease appears to have first been recorded in a legal text that mandated heavy fines for dog owners who did not keep a rabid animal under their control. That code was published in Mesopotamia four thousand years ago – making rabies one of the oldest regulated plagues on humanity.

Once in the brain, the first symptoms are seen as the parasite unleashes a three-pronged attack in its attempt to spread to another host. Viral bodies flood the salivary glands, paralysing the throat muscles. The saliva cannot be swallowed and spills out over the lips – the familiar "frothing at the mouth" of a rabid animal. In dogs, the paralysed throat muscles also give rise to the dreaded *voix au coq*, a rasping growl like that of a cockerel. The inability to swallow prevents the victim from drinking water despite an excruciating thirst. When attempts are made to administer fluids orally, the infected begin to panic, which is why rabies is also known as hydrophobia, a "fear of water".

So far, none of this is terribly unusual. Many illnesses – chicken pox, influenza, cholera, Ebola – will flood the body's fluids with an infectious agent and devise some way of presenting those to nearby hosts. Sneezing, coughing, oozing sores, vomiting and diarrhoea, spurting up blood.

What makes rabies unique – and more vicious in its way than Ebola – is how it takes control of its host's brain and commands it to spread the virus's offspring through contagious saliva. The disease does not simply stop the body's ability to swallow; it causes the host to become hyperaggressive. Think of the horror of Stephen King's *Cujo*. Indeed, rabies is derived from the Latin word for "rage", and *Lyssavirus* owes its name to the Greek *lyssa*, meaning "violence". The rabies virus lives up to its linguistic pedigree. Fuelled by parasites hacking into the brain, any infected animal succumbs to the rage: dogs, cats, foxes, raccoons, bats and humans alike. Like "envenomed weapons", the teeth are coated in virion-laden saliva; any bite inflicted by a rabid animal will inject thousands of viral bodies into the next victim. During the final stages of the disease, the parasite inflicts irreversible damage to the brain, fatally compromising its ability to regulate essential life functions, including respiration. Thus, once symptoms begin to manifest, rabies is always fatal. There is no cure. Crude, but effective: 55,000 people die from rabies each year. Naturally, this makes rabies the choice disease on which to model a zombie epidemic.*

Most of us will be lucky enough that we never meet a rabid animal, and so it's difficult to appreciate how ferocious an infected animal can be. The doctor Samuel Cooper, writing in 1822, noted that twenty-three people were bitten by a single rabid she-wolf. Half of the wolf's human victims died. The unlucky ones all received bites to their naked skin. Wolf bites were found to be more frequently infectious than those of dogs, because of the wolf's tendency to attack the face. The agony and inevitable lethality of rabies gave the disease a dreadful reputation; anyone could be bitten by a stray dog in the street, and once they had been, hope was lost. Sometimes people bitten by a rabid animal would commit suicide rather than face a slow, awful death. Outbreaks sparked mass panics. During the 1750s, the terror of rabies was so high that any dog found wandering the streets of London could be shot for a reward. Culls took place in other European cities, too: two thousand dogs killed during an outbreak in Hamburg, nine hundred killed in a single day in Madrid. Reducing the canine population, either by culling

---

* Horror flicks such as *The Crazies*, *I Drink Your Blood*, *28 Days Later*, *Rabid*, and *[Rec]* all feature infectious agents that bear some resemblance to rabies, although for the purposes of film-making, the long incubation period is slimmed down to just enough minutes to finish your vat of popcorn.

or by levying a tax on keeping animals, was the only effective tactic in controlling the disease.

Various folk remedies existed, such as swallowing a hair from the dog's tail – a literal hair of the dog that bit you. A sixteenth-century charm was to seize the rabid animal by the scruff of the neck and shout an incantation into its throat: "*Y ran quiran cafram, cafratrem cafratrosque!*". Those who didn't want to get that close to the jaws of a rabid animal could write the spell on a piece of paper, mix it into an omelette and feed the eggs to the dog. More sophisticated doctors suggested cleaning and sanitizing the wound immediately, advice still put forward today.

By the nineteenth century, the fear of rabies had boiled into outright hysteria. In 1885, the famed microbiologist Louis Pasteur developed a vaccine using a weakened form of rabies that he had isolated from rabbits. Without licence or ethical approval, Pasteur administered the experimental vaccine to nine-year-old Joseph Meister, who had been savaged by a rabid dog. Young Joseph survived, and went on to become caretaker of the Pasteur Institute in Paris.

That vaccine and its later versions are only effective as a prophylactic, administered before or within a few days of exposure to rabies – if symptoms have already developed, it is useless, which was bad news for fifteen-year-old Jeanna Giese. In autumn 2004, while warming up for a volleyball match at her Wisconsin high school, Jeanna was struck down by fatigue, fever, double vision, vomiting and pins and needles in her left arm. She was admitted to the nearby St Agnes Hospital, where her condition worsened. There she was put through a battery of tests, testing for everything from meningitis to Lyme disease; all returned negative results. "My left arm was jerking, I couldn't stand, I had excess saliva and couldn't talk", Jeanna reported. She was transferred to the Children's Hospital of Wisconsin to wait out the course of her mysterious illness.

Then her mother recalled that, three weeks earlier, Jeanna had been bitten by a bat that had flown into their parish church in Fond du Lac. At the time, Jeanna had rinsed the wound and thought nothing of it. "Those thirty seconds with the bat's fang in my finger would change my life forever", she said. The doctors checked Jeanna's blood and confirmed that it was packed with antibodies against *Lyssavirus*. To say her prognosis was poor was an understatement – the medical team gave her four hours to live.

One doctor at the hospital, Rodney Willoughby Jr, was not prepared to give up on Jeanna. He thought that she might have a chance of defeating the disease if her brain could be protected – just long enough to allow her immune system to mount a defence. Willoughby embarked on a risky and experimental procedure, inducing a coma with a cocktail of sedatives including ketamine, midazolam and phenobarbitol, then administered some antiviral medications, ribavirin and amantadine. Willoughby's goal wasn't just to put Jeanna to sleep, but to suppress all of her brain's activity, carrying her perilously close to the point of brain death, and potentially beyond it. He told Jeanna's parents that while her body could survive the treatment, her brain might not.

After a week, tests showed that Jeanna's body was fighting off the virus, and Willoughby and the team carefully eased her off the sedatives, awakening Jeanna from her coma. Jeanna had survived, but at a cost: she could not speak or walk. After months of intense rehabilitation, she slowly put her life back together, relearning basic life skills. Eventually, she earned her driving licence and enrolled at nearby Marian University, where she studied biology. Jeanna doesn't play as much sport as she used to, and jokes that she lists to one side when she runs. She had beaten late-stage rabies without protection from Pasteur's vaccine – the first person in recorded history to do so.

Dr Willoughby's experimental treatment is now known as the Milwaukee Protocol. How it works is not clearly understood. So far the protocol has saved four further victims of acute rabies infection, people who would have otherwise been felled by the virus. Still, rabies remains incredibly deadly: thirty-one others receiving Willoughby's treatment in various forms have not survived.

If rabies is the closest thing we have to our zombie virus, it seems fitting that a spell of death would turn out to be along the road to a cure.

A FATAL GAME OF CAT AND MOUSE

In the late afternoon sun of the Turkish summer, a middle-aged man drives through downtown Manisa, a city in the western reaches of the country, near the Aegean Sea. The weather is hot and dry, and the windows of his white saloon are rolled down to admit a cool rush of air. He weaves nimbly from one lane to the other in a wide road devoid of markings. At an intersection, the green traffic light begins to change to amber, but he can make it through, he thinks. He passes a high-sided van, already standing at a stop, as he sails into the junction. Too late, he sees the twin lanes of traffic approaching from his left, already well across the road. His speeding car is clipped at the tail end and flies into an uncontrolled spin, the back end swinging wide and nearly knocking down a pedestrian. A few seconds later, the car smashes into a wall, and the man becomes another one of the seven thousand individuals that make up Turkey's grim annual statistics of road traffic fatalities.

Most often, the blame for these accidents is placed on driver error – men and women driving too fast, too tired, too distracted. But what if there was something else at work? Maybe the accident – and hundreds more like it – had been planned in advance. What if the driver had been nudged through the red light, egged on by an unseen puppet master? After some hunting, Professor Kor Yereli of Celal Bayar University, Manisa, discovered that many of Turkey's road accidents were in fact being orchestrated by parasites lodged in the brains of its citizens. A murder-making epidemic was playing out on the streets, driving people to mortal distraction.

Yereli and his team took blood samples from a group of 185 men and women involved in traffic accidents in which alcohol did not play a factor. One-third of the people tested positive for antibodies associated with infection by a microbe known as *Toxoplasma gondii*, compared to a modest 9 percent in a similarly sized control group. Although they showed none of the symptoms normally associated with infection, those harbouring *T. gondii* were significantly more likely to be involved in a car crash. It was as if the parasites were interfering with their hosts' ability to drive – not enough to be noticeable, but enough to increase their chance of an accident, sooner or later. What can we discover from the wiles of this deadly intruder? How is it taking over its host – and more importantly, why?

Toxoplasma gondii is one of the world's most successful organisms. A tiny microbe measuring no more than ten-millionths of a metre across, it is found across the world, particularly in warmer climates. It takes its name from the gundi: a cute, stocky, somewhat shrivelled rodent, native to the arid plains of North Africa, which looks a bit like a mix of a hamster, a cat and a very furry prune. It was in a gundi that *T. gondii* was first identified by Charles Nicolle and Louis Manceaux in 1908. That same year, an Italian bacteriologist, Alfonso Splendore, found the parasite in some Brazilian rabbits. He was the first to describe the illness resulting from *T. gondii* infection, which, he said, bore a resemblance to the human disease *kala-azar* – Hindi for "black fever".* Splendore predicted that toxoplasmosis would likely also affect humans, though he did not yet have proof of it. "We should not be surprised if this disease would be observed in humans in the future", he told a symposium of doctors in Paris in 1912.

Several years later, studies confirmed that large segments of the human population harboured the bug, and that infection could result in miscarriages in pregnant women, making it a huge public health threat. Because of the risk of miscarriage, acute toxoplasmosis has been closely studied. Outbreaks can be particularly severe in flock animals such as sheep, where the parasite can trigger "abortion storms" with as many as half of ewes miscarrying. In the UK, about 22 percent of people test

---

* Black fever is a common name for *leishmaniasis*, a particularly nasty disease of the equatorial regions that is spread by sandflies. A few weeks after a sandfly bite, which involves bloodsucking, of course, a sore or ulcer forms – and can take up to a year to heal. The infection can even start to resemble leprosy. In black fever, its most severe form, it migrates to the vital organs, and almost always ends in death. The *Leishmania* genus ranks second among parasite killers, topped only by *Plasmodium*.

positive for antibodies to the protozoa, evidence that they have been exposed to the bug in the past. The rate varies wildly from country to country; in France up to 84 percent of the population tests positive. The difference is likely due to local eating habits – a society's taste for undercooked meats is correlated with a higher infection rate. To a much lesser extent, it is associated with proximity to cats, which is *T. gondii*'s primary host. It is this reason that doctors advise pregnant women to avoid contact with the family moggy.

Although scientists originally came across *T. gondii* in gundi and rabbits – and though it can infect itself into just about any mammal – the parasite can only complete its life cycle in a feline host. This has helped to make *T. gondii* so prevalent: it depends on a family of animals that has been domesticated, bred, nurtured and even worshipped as a hunter and friend in human households the world over – a resident of more human homes than "man's best friend", the dog. *T. gondii* can amble its way to another mammal in order to survive until the next cat walks by. It's so flexible about its living arrangements that it has even managed to establish colonies in dolphins, courtesy of cat owners who have flushed their pet's litter down the toilet, whereupon the animalcules journey through sewers and rivers, until landing in the ocean.*

There's one type of mammal that comes into close contact with cat carriers nearly as often as people: rodents. About 35 percent of the wild rat population is infected with *T. gondii*, and rodents are also key to its spreading around the globe. When

---

* *Toxoplasma gondii* is able to complete its life cycle in wild and large felines, too, but how often do you rub up against a leopard?

T. gondii is in a non-feline host, it can only "clone" itself, reproducing asexually by copying itself over and over in two feeding forms: the moving tachyzoite, which quickly fills up cells until they burst and are easy for the immune system to attack, and the stationary bradyzoite, which grows more slowly and hides inside little bubbles called vacuoles inside cells, waiting to jump from the host to a feline and gain its chance at sexual reproduction. The bradyzoites in the rats can be transmitted from mother to offspring, but they are also topped up by fresh batches of oocysts – tiny eggs packed with super-infectious sporozoites – found in cat droppings. It's not that the rats are frolicking in cat faeces (at least not regularly); the droppings often mingle with food scraps in our rubbish, contaminating the rats' dinners. After a rat inadvertently ingests *T. gondii*, generations of tachyzoites plunder the body's cells, until some of these parasites convert into bradyzoites and settle into the brain, liver and muscle tissue, where they hide in their protective cysts. To break free of this plane of existence, the parasite must find a cat host – or, in other words, a cat must catch a rat for its own supper.

Parasites have developed a number of ways to spread from one host to another – hitching a ride on a passing vector, such as a mosquito or a flea, or dropping out of the host and hoping that another suitable one will stumble across it, like the ant-infesting *Ophiocordyceps unilateralis*. *T. gondii* are more cunning than that. After going through generation after generation of sexless existence, it is in no mood to wait on chance encounters. Instead, *T. gondii* reprograms the rodent's brain.

Rats are generally cautious animals. Their eyesight is rather poor, and they stay away from the open spaces and bright light that make them vulnerable to predators such as dogs, raptors and cats. Rats use their excellent sense of smell and sensitive whiskers to remain close to cover, laying down scent trails and sticking to well-known, well-travelled and safe routes. As you might expect, they possess an instinctual fear of cats, and will avoid an area if they catch even the slightest whiff of cat urine. This behaviour is so deeply stamped in the rodent brain that even laboratory rats that have not encountered a cat for five thousand generations show an aversion, growing anxious when they detect the smell of a nearby feline. This poses something of a problem for *T. gondii*, because only sick or very unlucky rats are likely to end up on a cat's supper plate. The parasite might try inflicting some kind of physical damage on its host, to slow it down and make it an easier target, but a sick rat is just as likely to die in its burrow or at the claws of other rodents than it is to be killed by a cat. So *T. gondii* takes another tack: it boosts the rat's ego. Manuel Berdoy of Oxford University calls it a rodent form of "fatal attraction".

Berdoy and his colleagues revealed the effect of *T. gondii* on rat behaviour in a study published in the year 2000. The scientists captured rats from British farms and crossed them with laboratory specimens, producing half-wild animals that combined natural behaviour with known histories. After a brief course of antibiotics to remove any other parasites that may have been lurking in their bodies, the half-breed rats were infected with *T. gondii*. They were then placed in a two-metre-square construction, which was subdivided by brick mazes into sixteen smaller rooms for the rats to explore. In each corner of the construction, a small den was set up with food and water and doused with a particular smell. One was kitted out with clean straw (for a "neutral" smell), while another used bedding taken

from the rat's "home" den to create a familiar and cosy environment. The two remaining corners were sprayed with the urine of
other mammals – the enemy cat and the less-threatening rabbit,
which is not a predator of rodents. The pens were cleaned and
the positions of the four dens were shifted daily. Which corner
would the rats prefer?

Rats are nocturnal, so each night the team would let them
explore their new environment (on camera, *Big Brother*-style).
The uninfected rats acted as rats normally do – avoiding the
den laced with cat urine and staying mostly near their own
bedding. The infected rats also showed a preference for their
own smell, but they exhibited less hesitation about inspecting
the den thick with pheromones from a cat. In fact, over the
course of the study, the infected rats preferred to visit the den
smelling of cat over the one smelling of rabbit. The parasite
wasn't just suppressing the rat's sense of smell – it was actively
reprogramming its response to olfactory cues, willing it to be
attracted to its natural enemy. Infected rats also visited more of
the maze's cells than their uninfected cousins, exposing themselves to unknown environs. They appeared more curious and
more willing to put themselves at risk. As Dr Joanne Webster,
one of the members of Berdoy's team, put it: "The infection
makes the rats less fearful of novelty. Rats can usually detect
subtle changes in their environment. It makes them very hard
to trap or poison but this parasite overrides the innate response;
they almost taunt the cats in a sense."

This may seem like an inconsequential game of cat and
mouse – but recall the risky drivers on the streets of Turkey
(and likely near you too). Crucially, humans and other warm-
blooded hosts can be colonized by a small population of *T.*

gondii, whereby we do not suffer any clinical symptoms and the parasitic infection goes unnoticed. Then, if the immune system is compromised, as with individuals undergoing chemotherapy or suffering the late stages of AIDS, the infection rears its ugly head. Unfortunately, in these cases *T. gondii* doesn't make us too curious for our own good; rather the infection leads to dementia, with *Toxoplasma* punching "great big holes in your brain" (to use Webster's words). "I don't think we can be that dismissive of such a prevalent parasite in our brains", she cautioned.

It would not take long for evidence to surface in support of Webster's warning. When Berdoy and his team released their results, Jaroslav Flegr, a professor at Charles University in Prague, had been hunting for over a decade for the spell that *T. gondii* might have over humans. Over the next several years, Flegr continued to probe the link between *T. gondii* and human psychology, finally publishing a summary of his studies in *Schizophrenia Bulletin* in 2007 (his choice of journal may hint at his conclusions).

Curiously, Flegr found that the sexes were not equally disturbed by *T. gondii*. Men with the parasite show lower superego strength (the part of the psyche that restrains selfish impulses and allows us to act socially) combined with higher vigilance – meaning they become more rash-minded and more likely to disregard rules, as well as "more expedient, suspicious, jealous, and dogmatic", when *T. gondii* enters their brain. Women, on the other hand, display more warmth and higher superego strength; the parasite makes them "more warm-hearted, outgoing, conscientious, persistent and moralistic". Unlike in the rats, levels of apprehension grow in all infected humans, regardless of their sex.

Previous animal studies had shown that infection with *T. gondii* impaired motor performance, and Flegr wanted to see if that was also true in humans. Flegr and his colleagues asked 120 adults, half of whom tested positive for antibodies to *T. gondii*, to take part in a test of reaction time. When a white square appeared on a screen in front of the subjects, they were told to click on a button – a simple task. Those carrying *T. gondii* in their bodies performed significantly worse than "clean" individuals. The finding has been replicated in two subsequent and much larger studies, though the results have not yet been published.

Could such differences, quantified by psychological profiling and reaction tests, translate into real-world, life-and-death changes in human behaviour? By these measures, *T. gondii* clearly affects people, but so do innumerable other factors – tiredness, fitness, overall health, mood, age, upbringing and genetics. Could *T. gondii* really warp our brains to kill us off in the same way that it leads rats to their death? Flegr thought so. His next stop was one of Prague's hospitals, where he collected samples from 146 people injured in road traffic accidents in which they were deemed to have been at fault, whether as drivers or as pedestrians. Compared to samples collected from a random cross-section of the Czech public, he discovered that those infected with *T. gondii* were over two and a half times more likely to have been involved.

PROGRAMMED TO KILL

Some might feel a little piqued by the idea that an organism no bigger than a red blood cell can get you killed. After all, orchestrating an untimely death inside a speeding metal box serves no purpose to a host-embedded parasite, and spells its own demise as well. So what is the evolutionary point? Humans are biological dead-ends as far as *T. gondii* is concerned, so why should it infect us? One suggestion is that we're not too unlike rats – that the parasite doesn't recognize or care that it's in the wrong host, and employs its mind-altering tricks regardless. If we wanted to speculate wildly, we might say that the answer lies back in the days when our ape ancestors had only just started to climb down out of the trees. At this time, the forest floor was a dangerous place, with big cats stalking us from the ground and from the trees. Primates were (and still are) a major source of food for feline species, and *T. gondii*'s manipulations would have helped to ensure that the parasite could jump back into a cat and reproduce sexually. In modern-day Prague and Manisa, *T. gondii* keeps up its efforts to get us killed, though the only big cat we're likely to fall victim to is a Ford Puma.

But how does *T. gondii* pull off its tricks? That's the question that's been puzzling neurobiologists since Flegr and Webster conclusively demonstrated the parasite's mind-altering abilities. We know that *T. gondii* is very exacting in the damage it causes to the brain's circuitry: the infected rats in the maze study showed no signs of broad-range neurological problems, and they responded normally to other fearful stimuli, including bright lights and open spaces. *T. gondii* did not operate simply by injecting chaotic storms of neurotransmitters into the brain to befuddle the rats; rather, it precisely rewired certain parts of the brain without interrupting normal function in the rest of the cells.

It turns out that "rewiring" is a surprisingly appropriate turn of phrase – it's not just a computer-era metaphor for the functioning of the brain. When *Toxoplasma* spores are ingested by a rat, they are absorbed through the gut and begin their slow migration to the brain. After about six weeks, the bradyzoites are concentrated in the amygdala, the part of the brain responsible for conditioning appropriate fear responses. In infected rats, the long branching dendrites that connect nerve cells to one another shrivel up – *T. gondii* physically severs the amygdala's circuits. Somehow, the parasite is able to hone in on the specific circuit relating to fear of predators, disabling that whilst leaving other fear pathways intact. How it is able to identify this particular pathway is still unknown, but it makes our best neurosurgery look like trepanation carried out with a blunt chisel – not even an ice pick. And this is only half the story.

Removing the millennia-old fear of cats in rodents is one thing, but creating an attraction to their pheromones requires hacking another neural pathway. *T. gondii* rewires the parts of the brain responsible for sexual attraction, so that the smell of cats triggers the reward pathway that is usually triggered by the presence of a potential mate. In an interview with the online science salon *Edge*, Robert Sapolsky, a professor of biological sciences at Stanford University School of Medicine, discussed the nuts and bolts of this uncanny trick:

> Toxo [*T. gondii*] knows how to hijack the sexual reward pathway. And you get males infected with Toxo and expose them to a lot of the cat pheromones, and their testes get bigger. Somehow, this damn parasite knows how to make cat urine smell sexually arousing… This is utterly bizarre.

Yes, that's right: *T. gondii* makes rats *sexually* attracted to cats. Fatal attraction, indeed.

Sapolsky reveals that the parasite is able to pull off its trick because it possesses something rather unique: the mammalian gene that codes for the manufacture of dopamine. Dopamine is a neurotransmitter responsible for a huge variety of reward behaviours in humans – it's what makes drugs, sex, chocolate and a host of other pleasures pleasurable. *T. gondii* doesn't need this dopamine for itself – in fact, it only starts producing dopamine once it has migrated into the host's brain. It produces the chemical with a little tail that extends into the cell in which it's hiding, and instructs that cell to excrete the neurotransmitter into the immediate environment – the brain. With dopamine at its command, *T. gondii* can design new pleasure-reward pathways, seducing its host into enacting the "unnatural". Predictably, this talent has intrigued certain organizations. According to Sapolsky:

> You want to know something utterly terrifying?… Folks who know about Toxo and its effect on behaviour are in the U.S. military. They're interested in Toxo. They're officially intrigued. And I would think they would be intrigued, studying a parasite that makes mammals perhaps do things that everything in their fibre normally tells them not to because it's dangerous and ridiculous and stupid and don't do it. But suddenly with this parasite on board, the mammal is a little bit more likely to go and do it.

The military brass, as well as those who think that a spoonful of *T. gondii* oocysts might be just the thing to overcome a fear of heights or public speaking or, dare I say it, cats, might want to think twice, however, about dosing with the parasite. In humans, excessive dopamine production is associated with mental illnesses such as schizophrenia – which was part of the reason Flegr's study was of interest to the editors of *Schizophrenia Bulletin*.

Reasoning that *T. gondii* might also control rat minds by altering dopamine levels, Joanne Webster ran an experiment in which she treated infected rats with haloperidol, a drug often used in the treatment of schizophrenia. Haloperidol works by inhibiting dopamine receptors in the brain. If her hypothesis was correct, desensitizing the rat's brains to extra dopamine in this way should render *T. gondii* impotent. The rats fed on the antipsychotic drug regained their natural aversion to the smell of cats, demonstrating a significant reduction in suicidal behaviour. She was right.

But this finding could also be turned on its head. If *T. gondii* causes excessive dopamine production, responds to antipsychotics, and is known to infect humans, might *T. gondii* itself be the underlying root cause of schizophrenia? A correlation between toxoplasmosis and schizophrenia had been noted as far back as the 1950s, but the link was fairly tenuous – for example, doctors noticed that mothers who kept cats during pregnancy were more likely to be diagnosed with schizophrenia. In 2003, two researchers based in Maryland, Fuller Torrey of the Stanley Medical Research Institute and Bob Yolken of Johns Hopkins University, discovered that those suffering from schizophrenia were almost three times as likely to test positive for *T. gondii* antibodies than the general population. In a separate study, Yolken found that patients suffering from mood disorders were more likely to attempt suicide if they carried a very high load of *T.*

gondii antibodies. This research doesn't verify that *T. gondii* causes mental illness, but it is under suspicion of playing some role.

That was enough of a suggestion to inspire Teshome Shibre, an assistant professor of psychology at the Aklilu Lemma Institute of Pathobiology in Addis Ababa. He decided to run Webster's experiment in reverse, to see whether *T. gondii* might be the key to a cure for schizophrenia. If the microbe played a leading role in the illness, those who suffer from it might be cured by giving them a course of cheap antibiotics. To test this hypothesis, Shibre and his team asked 159 patients, all of whom were diagnosed with moderate to severe schizophrenia, to take a drug for six months. Half were given the antibiotic trimethoprim, which inhibits bacterial growth and is typically used in the treatment of urinary tract infections. The remainder of the group were given a placebo. At the end of the six-month period, the group on trimethoprim showed a significant reduction in their schizophrenic symptoms – but so did those on the placebo. In other words, there is no evidence of a "magic bullet".

It may be that the damage wrought by *T. gondii* is permanent and debilitating for some individuals. It may be that the microbe acts as a trigger for the disease, paving the way for some other pathogen to unleash itself in the body and develop the symptoms we see. Or it may be that the microbe and the disease are completely unrelated, nothing more than coincidence. The answers are still out there, waiting for a scientist to discover them.

So, bad news for anyone who is hoping to enslave the world with their own zombie plague – the natural world has beaten you to it. Even the US Centers for Disease Control have issued guidelines on what to do in the event of our imminent zombie outbreak, with Director Dr Ali Khan setting out the gravity of the situation: "If you are generally well equipped to deal with a zombie apocalypse you will be prepared for a hurricane, pandemic, earthquake, or terrorist attack."

The fact is, tiny organisms like *T. gondii* are not just skilled at hacking the minds of their hosts, they have a robust and widespread distribution network established. The rise of the zombies is happening all around us, if we'd only take a moment to notice.

Now, what could we do with our human zombies, once we find a way to master their process?

7

THE HUMAN HARVEST

Death is the only thing we haven't
succeeded in completely vulgarizing.

Aldous Huxley, *Eyeless in Gaza* (1936)

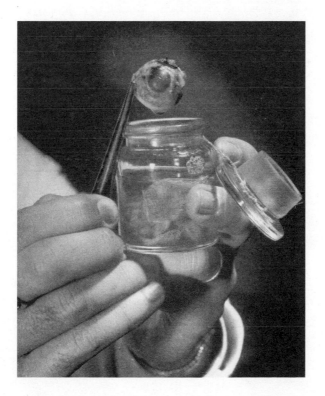

WHEN THE FAMOUS PROGRAMMER SID MEIER
created *Civilization*, the wildly popular video game that gives
you a chance to guide the development of a band of humans
from their Neolithic scrambling to a space-faring destiny, he
chose to have ceremonial burial offered alongside masonry and
the wheel as one of the first fundamental "technologies" that a
player can discover.* Once you have acquired death rites, you
can delve into mysticism and philosophy, and thereafter build
an empire marked by the possession of organized religion,
astronomy, mathematics, physics and so on.

Civilization might be just a game, but Meier was right:
our relationship with the dead is a cultural cornerstone and a
stumbling block. Ceremonial burial neatly captures both our
idealization and experience of death: for all the hymns and
prayers dedicated to the immortal soul, you still want to burn
or bury the corpse before it starts to go soft and rot. For all of
our technological advances (and often, because of them), we still
wrestle with the pragmatic issue of what we do with the body
once it's dead. On the one hand, the dead are often laid to rest
after a series of solemn rites made to honour a person's accom-
plishments and character in life; sometimes bodies are stashed

---

* The earliest undisputed burial of a human corpse, discovered
in a cave in Israel, dates to between 80,000 and 130,000 years ago –
Palaeolithic rather than Neolithic in its roots. But playing a video
game starting in 130,000 BCE (rather than 4100 BCE) would probably
take too much time.

inside grandiose monuments, locked up to protect them from the ravages of nature and the insults of foes and fortune hunters, and made into a tourist attraction for generations to come. On the other hand, we have a very long history of exploiting the human corpse as a handy resource, raiding it for medicines, spare parts and even food.

As we stare down our real-life space-faring destiny, it's only natural that the dead remain on our minds. Just a few years after Neil Armstrong planted his boots on the moon, Dr S.L. Henderson Smith made headlines for a radical proposal for handling Britain's dead. The "ecological" alternative to burial in a casket – cremation – was energy intensive and wasteful, a "gross waste of valuable resources", he said. Corpses, he suggested in his letter to the journal *World Medicine* in 1973, could be put to far better use as, say, fertilizer. The *Guardian* got a whiff of the story, explaining that Smith "could see the day when a body alone was either reprocessed as fertiliser or mixed with sewage for the same end. It might even provide a new form of fuel." Smith wasn't the only British visionary to imagine recycling the dead into something productive – but then, Aldous Huxley had his tongue nestled firmly in cheek when he dreamt up crematoriums that extracted phosphorous from the corpses of London, AD 2540, in his novel *Brave New World*. Predictably, Smith's idea raised an outcry, and the indomitable public moralizer Mary Whitehouse was on hand to blast the proposals as "a pre-civilization concept… subhuman, let alone subspiritual".

Smith was destined to join the legions of well-intentioned scientists who have misjudged the depths to which sentimental values hold firm – or perhaps you could say he was simply ahead of his time. Environmentally friendly burials are now on the rise, as is donating your body for scientific or medical use; mulching your loved ones and blending them with sewage has been slower to catch on, but there are outfits that are poised to fulfil your needs. Resomation, based in Glasgow, Scotland, offers alkaline hydrolysation, which often goes by the less precise but more catchy term "liquefaction". Alkaline hydrolysis is touted as an environmentally friendly alternative to cremation as it consumes less energy and produces fewer toxins. After death, the body is interred in a giant metal tube. The interior of this gleaming sarcophagus is then filled with boiling lye and held at ten times atmospheric pressure. The remains quickly break down over the next few hours. The sterile liquid waste is poured down the drain, and the crumbling bones are ground to "ash" and returned to the bereaved in an urn.

Another company, Aquamation Industries, with offices in Australia and the US, has been launched to adapt the "water cremation" equipment supplied to farmers for the disposal of animal carcasses as a replacement for cremation by fire. In August 2010, the company opened for business, and in the first few months claimed that 60 (living) people had signed up. The corporate website mentions that "another advantages [sic] of the Aquamation Industries Equipment, is that it actually has a vent to enable the soul to leave the body, and to go to Heaven" – a feature that is not available in "high pressure equipment", which seems to be a knock at rival Resomation.

In the agricultural industry, the resulting residue of the processes is known to make "a great fertilizer or additive to composting farm wastes", say the entrepreneurs. Neither company, however, has thus far drawn attention to this particular benefit as a selling point for their human funerals.

But that doesn't mean others aren't more than ready to put zombie flesh to work.

THAT GOOD-FOR-SOMETHING ZOMBIE

If you heard that customs officers had intercepted a shipment of illegal Chinese medicine, you might assume that the person was talking about supposedly aphrodisiac pills made from rhino horns or tiger genitals. South Korean agents did not have such an easy case on their hands when they seized over 17,000 capsules of Chinese "stamina boosters" made from the powdered flesh of human foetuses. The unfortunate donors, arising from miscarriage or abortion, had been collected by medical staff in clinics in the Jilin region of north-east China. The bodies were then passed on to herbal shops, where they were processed and sold as health preparations.

Lest we be too quick to label these baby-pill poppers as barbaric, we ought to remember that beliefs about the restorative power of human bodies have been rife in Europe too. At least since the twelfth century, bitumen was widely held to be a natural tonic, and thought to be the preservative used by ancient priests to embalm the dead. Dusty fragments of the shrouds surrounding Egyptian mummies were much sought after for their supposed medicinal properties. The fact that Egypt's desiccated dead had been preserved with resin-soaked cloths, not bitumen, did little to diminish this fabled power. By association, the tonic's "healing power" gradually leached from the cloths into the cadavers they swaddled, and in time it was the mummified bodies themselves that were coveted for a range of medical preparations. Ambroise Paré, the sixteenth-century pioneer of battlefield surgery, noted that this corpse liquor was "the very first and last medicine of almost all our practitioners' against bruising".

The popular preference, though, was for products derived from more robust humans. A common theory in Europe was that each person had a set lifespan, and that the bodies of those who died before their time, due to accident or violence, retained in them a surplus of life energy, which could be extracted and added to another individual's supply. Following this logic, the ancient Romans would drink the blood of gladiators to share in their intrinsic vital essence; it was believed to be especially effective at curing epilepsy. This idea persisted long after gladiatorial combat went out of fashion, and so people were forced to find other sources for blood. So prevalent was the belief in the curative power of prematurely spilled blood that people would jostle for position in the crowds attending the execution of criminals. The closer you were to the platform, the better your chance to catch any droplets of blood raining down. Spectators would hold out cups; they could also pay a small sum to mount the platform after the execution and dip a handkerchief into the puddles of blood, and it was believed that wiping a fresh wound with the cloth would speed up the healing process. In 2008 one such handkerchief, stained with the blood of a personage no less vainglorious than Charles I, sold at auction in Swindon, Wiltshire, for £3,700 (about $6,500) – though it was probably valued as a historical curio rather than a health aid.

Such vitalist beliefs persisted up to the eighteenth century, meaning that for several hundred years medicines made from human remains were a regular feature in a European surgeon's toolkit. A sixteenth- or seventeenth-century patient could expect to come across dozens of these compounds, according to the medical historian Richard Sugg, who details some examples in his bluntly titled article, "'Good Physic but Bad Food': Early Modern Attitudes to Medicinal Cannibalism and Its Suppliers". The Danes' predilection for blood, he notes, was shared by many of their neighbours, and was being recommended as a treatment for epilepsy by English physicians as late as 1747 (the blood was best served "recent and hot"). Hans Christian Andersen, in his autobiography, reported the 1823 case of a boy whose superstitious parents forced him to drink a cup of blood from an executed criminal, in the hopes it would cure the child's epilepsy.

Blood was not the only curative. Human fat was considered good for the treatment of rheumatism and arthritis, like applying oil on a creaky hinge. And King Charles II paid a fortune for a formula revealing how to distil a human skull – another popular ingredient in healing tinctures. Because of this royal patronage, the resulting elixir became known as "the king's drops".

Towards the end of the eighteenth century, doctors in Europe shied away from using tissue from humans, dead or alive, in medicines. It wouldn't be too long, however, before advances in medical technology reignited the question of whether it was possible, or appropriate, to extract restorative powers from the dead, especially from their lifeblood.

In 1812, the French physiologist Julien Jean César Legallois predicted, "If one could substitute for the heart a kind of injection... of arterial blood, either natural or artificially made... one would succeed easily in maintaining alive indefinitely any part of the body." In the decades that followed, many attempted to do just that, with (as we saw) increasing success, especially between the World Wars. One of the pioneers was Dr Alexis Carrel, who in 1912 was awarded a Nobel Prize for developing techniques to suture major blood vessels and perform complex grafts. He also helped to make breakthroughs in the use of antiseptics, particularly chlorine, in surgeries during the First World War, for which he was awarded the renowned French Legion of Honour.

Carrel's early success in grafting blood vessels laid the groundwork for an even more ambitious goal: organ transplant. Carrel worked tirelessly at the Rockefeller Institute for Medical Research, in New York, perfecting the techniques necessary to keep organs alive outside the human body. Much like his Soviet contemporary Sergei Bryukhonenko, Carrel pinpointed that nutrients were essential for keeping the organs healthy enough to survive for more than a few moments of study. He created a heady cocktail of blood serum, insulin, thyroxine (a thyroid hormone), vitamin A, vitamin C and more, and developed an apparatus to perfuse blood with his cocktail so that it could be circulated through an extracted organ. But no matter how carefully he cleaned his equipment, his pump always introduced contaminants, and infection always set in. What he needed was an entirely new type of pump, an artificial heart that could hold the blood in complete isolation from the outside world and transfuse it into the organ as though the system were a discrete organism.

By a twist of fate, Carrel was introduced to the aviator Charles Augustus Lindbergh – the same Charles Lindbergh who had just made the world's first non-stop flight from New York City to Paris. The two became close friends, not least because they shared common political and social views. Carrel, like Lindbergh, was a strong supporter of eugenics, and was vocal in his critique of those he felt stood in the way of human progress. "There is no escaping the fact that men are not created equal, as democracy, invented in the eighteenth century – when there was no science to refute it – would have us believe", he argued in his bestselling book, *Man, the Unknown*. He went on to suggest that certain segments of society ought to be euthanized in gas chambers – the criminally insane, those who have murdered, taken up armed robbery, abducted children, even "those who have misled the public on important matters". For his part, Lindbergh wrote, "We can have peace and security only so long as we band together to preserve that most priceless possession, our inheritance of European blood, only so long as we guard ourselves against attack by foreign armies and dilution by foreign races."

As tends to be the case with eugenicists, both Carrel and Lindbergh felt they represented *the* ideal, which ought to be pursued at all cost. Their views aligned with the governing social and political theories of Germany at the time, and both men were branded as Nazi sympathizers. Neither seemed much to mind. Lindbergh openly expressed admiration for the Nazis for pursuing social policies that he felt were most pressing. And in Nazi-occupied France, Carrel found a welcoming home for his long-desired Foundation for the Study of Human Problems, where he oversaw the implementation of various eugenics policies.

When Lindbergh first visited Carrel's lab in New York, he was astonished by the equipment on show. He could not believe that a great medical mind should be hampered by primitive engineering. He offered to design (anonymously) a superior blood pump for the Frenchman, and he came up with an ingenious solution to the problem of infection of the blood supply.

Lindbergh's pump consisted of an elegantly spiralled glass tube wrapped around a straight one. When the device was rocked back and forth, centrifugal force would push liquid to the top of the spiral, from whence it would splash back down the vertical tube to the bottom. It wasn't perfect, but the airtight device allowed fresh oxygen and nutrients to be injected carefully through tiny valves. Lindbergh called it his "glass heart".

Carrel could now keep organs alive for much longer than was previously possible. He soon set about liberating "hearts, kidneys, ovaries, adrenal glands, thyroid glands, spleens" from chloroformed cats and chickens. Placed in an aseptic container and connected to the artificial heart, the organs thrived and even grew. In one instance, an ovary swelled from 90 milligrams to 284 milligrams over the course of five days, and even seemed to produce eggs, suggesting that, with further tinkering, the device might one day be able to fertilize and incubate motherless eggs – the equivalent of an artificial womb. "An unrestrained imagination," one reporter wrote breathlessly, "could foresee Drs Carrel & Lindbergh placing whole animals – chickens, cats, dogs, possibly superannuated human beings – in their wobble machine and keeping them alive indefinitely." This unrestrained journalistic imagination was in no small part perpetuated by Carrel's erroneous belief that cells could continue to grow indefinitely. For twenty years Carrel kept a flask in which he claimed to have living cells from an embryonic chicken heart, a small demonstration of the immortality heralded by his transfusion work. It was an experiment that was never successfully replicated. What eugenics-inspired plans Carrel may have envisaged for the "outmoded" humans that made their way to his machine are left to *our* imagination.

Despite this showboating, Carrel's immediate aims were somewhat prosaic. He hoped Lindbergh's glass heart might one day allow him to grow organs such as endocrine glands, so that large quantities of hormones could be harvested in the lab. This would spare endocrinologists from the grim job of picking through carcasses at the local slaughterhouse, the usual method for getting their supplies at that time.

It's too bad that he didn't think to have them visit the local morgue instead.

IN COLD BLOOD

Around 1935, an American surgeon named Leonard L. Charpier embarked on a series of secret experiments at a hospital morgue in Chicago. He was looking for a solution to a tricky problem: blood was always in short supply.

In its early days, blood transfusion was very different to what we are familiar with today. Most transfusions were carried out using the "direct method", drawing blood from a donor through a tube that led directly into a vein of the recipient. This meant that if a surgical patient began to haemorrhage, the attendant physicians would have to storm out in search of a suitable donor, whisking that person off the street and into the operating theatre, without any time to put on sterile drapes. As well as being an incredibly time-consuming and chaotic process, the unscrubbed donor was a frightening source of contamination and infection. The donor was even a potential fire hazard if he or she happened to be wearing woollen or silk clothes, which might create a spark of static electricity and ignite an operating theatre filled with anaesthetic gasses.

Physicians were well versed in the use of sodium citrate for preserving blood once it was withdrawn from donors, so blood donation didn't need to be done on the fly. But there was no standard scheme in place for recruiting donors in advance of a patient's demand, and a healthy human could only comfortably donate one pint of blood every few months. Charpier realized there were people in the hospital who could afford to give more – a lot more. And so he began quietly drawing blood from the bodies in the morgue.

The practice of taking blood from cadavers had been pioneered by physicians working in – it almost goes without saying – Moscow. In 1930, a large-scale programme for harvesting blood from the dead had been established at the Sklifosovsky Institute, the capital's premier trauma centre. Over the following thirty years, the institute transplanted thirty tons of cadaver blood into living patients. When the first international conference on blood transfusion was held in Rome, Soviet scientists attempted to evangelize their methods, but the idea failed to spread. In the US, medics rejected the procedure, citing the potential risk of infection; with the immune system out of commission after death, the bacterial population in a body increases dramatically. In addition, early tests indicated that the Americans could not replicate the work of the Soviet scientists, who said they were able to wring up to eight pints of blood from a single cadaver.

Charpier decided to investigate whether there was any truth to the Soviet claims. A solidly built man, square-jawed and big-boned, he had played professional football with the Racine Cardinals while studying medicine at university in the 1920s, earning the nickname "Tank" for his formidable physique. Working as a doctor at the Little Company of Mary Hospital, located in one of Chicago's south suburbs, Charpier began his own experiments with cadaver transfusions, drawing blood out of an estimated thirty-five corpses over the next two years. With the assistance of medical intern Dr Donald F. Farmer, Charpier collected blood from bodies in the hospital morgue no more than four hours after death, and kept the drawn blood on the shelf for at least a week while samples were incubated and tested for contamination. Blood was only taken from males under fifty years of age, and only from those already undergoing autopsy, to avoid having to ask the relatives of the deceased for their permission. The recipients, likewise, had no idea that they were receiving blood from someone who was dead.

The top-secret project was an unqualified success. The Soviet scientists had indeed solved the problem of blood shortages – yet Charpier never shared his findings. That's because the cadaver-donor programme was almost immediately rendered obsolete by the development of a more American-friendly system: the blood bank.

The brainchild of Bernard Fantus, another Chicago doctor, the blood-bank concept took advantage of a metaphor that was extraordinarily easy for the public to relate to in 1937: a savings and loan bank.* In the early days of the campaign, a patient could make a deposit into the bank, creating an account against which a withdrawal could be made later if needed by the person, or by their family or friends. Hospitals even produced credit slips noting the amount deposited and any withdrawals. Fantus even

---

* Fantus's other entrepreneurial brainstorms included candy-coated medicine for children and "Make Chicago Sneezeless", a slogan for selling a hay-fever treatment.

suggested supplementing the bank's assets by approaching hypertensive patients – who might be coaxed into "paying in blood" in hopes of reducing their blood pressure (akin to bloodletting, the treatment advised by Galen and Hippocrates but no longer in favour). To build up supplies, a shareholding system soon developed, allowing patients to buy blood, the proceeds from which could then be paid out to donors, encouraging participation in the scheme. The Christian Men's Industrial League, a charity serving Chicago's homeless population, arranged for "unemployed and transient men" to be sent to the hospital to collect ten dollars in exchange for their blood. Fantus had applied the tenets of capitalism to the medical system of blood transfusion, and with great success.

Charpier's work only came to light in 1960, when his former assistant, Donald Farmer, who by then was director of the Beverly Blood Center in Chicago, published an article in the *Bulletin of the American Association of Blood Banks* about a cadaver blood programme in a local hospital. It was the only such programme ever established in the US – as far as anyone knew. Farmer's revelations came as a shock to two pathologists working at Pontiac General Hospital, in Michigan, who had just begun exploring the use of cadaver blood, unaware of the earlier research. The pathologists were treating a forty-one-year-old woman suffering with anaemia. After receiving a pint of blood from a twelve-year-old boy, her condition improved, and so she had been given a second infusion from the same donor. Of course, what allowed the same donor to give two pints of blood in such a short span was that the child had drowned in a lake two weeks earlier. The transfusion was the fourth that the doctors had done. Each time they had taken care to draw the blood from victims killed by accident or violence, not by disease, and to treat patients suffering only incurable diseases. These medical mavericks were named Glenn W. Bylsma and Jack Kevorkian – the same Jack Kevorkian who would later earn the nickname "Dr Death" for his outspoken campaigning to give terminally ill patients the right to die with the assistance of a medical professional.

BUTCHER, FAKER, BREAST IMPLANT MAKER

Kevorkian and Bylsma had been encouraged to study the value of the blood procedure after advances in other transplants, such as with corneas in the 1930s (in the Soviet Union, once again), had helped to dispel many of the prejudices against the use of material from dead donors. Over the subsequent fifty years, attitudes have softened even more, to the point where receiving a donation is accepted – the issue is finding organs.

Somewhere right now in the darkened room of a well-equipped hospital, a patient lies senseless, animated only by a ventilator, which steadily works air in and out of his tired lungs. Tubes and wires sprout from his body and weave their way into hissing, humming, beeping machines, which keep his blood flowing and maintain the right balance of critical electrolytes in his bloodstream. It takes a team of three nurses to watch over him, constantly checking his vital signs and carefully adjusting his medications. He is waiting patiently for a transplant to be carried out later in the day. Other than the mechanical chirping and the shuffling of the nurses, it's quiet. There are no visitors here to cheer him before his surgery. That's because the man's family have gone home to plan his funeral; he is dead. Our patient is not the recipient. He is the donor.

Despite Alexis Carrel's hopes, today most human organs can survive outside the body for only a matter of hours, and by far the most suitable storage vessel is the human body itself. Once removed from the warmth and comfort of the body, and despite the use of preservatives and ice baths, fragile tissues soon begin to degrade. By regulating the vital needs of a deceased person's body, medical staff can take the place of the brain and keep the body working. The body's breathing rate, normally the responsibility of a crinkled piece of brain stem known as the medulla oblongata, is now controlled by a ventilator-machine dial. Tiny adjustments to the blood's calcium and potassium levels help to maintain a steady heartbeat. Doping the blood with vasoconstrictors shrinks the vessels, keeping blood pressure up and pushing the blood to get oxygen to every bit of tissue. Such "donor maintenance" must be performed until a recipient is identified and prepared for transplant, exiling many of the dead to this twilight zone of functioning biology that few of us would call life.

Most people tend to view "brain death" as an ultimate, irreversible point at which a human passes away, but the term is ambiguous – more so today than at any previous moment in the history of medicine. First, you have to decide what you mean when you say that "the brain" is "dead". The term "brain death" can be used to refer to various types of injury to various parts of the brain. The human cerebrum, made up of the whorled grey matter of the neocortex and the almond-shaped basal ganglia, is believed to be the seat of human consciousness, of our memories and personality – what a practitioner of Vodou might call the *ti-bon anj* and what a Christian might call "the soul". Without it, a body might be seen as an empty vessel, or meat. If the medulla oblongata keeps functioning, running our basic metabolic processes, but the "higher brain" is irreparably damaged, how alive are we? If a machine can run all of the stuff that the medulla oblongata runs, does its "death" mean anything?

The viewpoint that the higher brain is what makes us not just alive but human has allowed many generations to embrace the concept that we are able to transcend death, that unlike most every other living thing, we do not decay but escape our bodies to exist on another plane – conveniently, in most versions of the story, untouched by age or disease. This separation of a person's body from his or her essence is partly responsible for the growing acceptance of organ donation in the past few decades: it allows us to believe that a person is immune from the insult involved in recycling organs – that Grandma remains whole, in an afterlife, or simply in our memories.

It's tempting to view organ donation as we see it on TV – beginning with an acute illness suffered by an endearing patient, then a desperate search for a donor, and the final conclusion that draws two lives together in bittersweet dénouement, one ending, the other starting anew. The reality is a far less intimate process from which few scriptwriters could extract any heartwarming scenes, for you or your Grandma. There are the waiting lists, of course, and the disappointment when it becomes clear that an organ will not make it to a person in time for lifesaving surgery. The US government has contracted with the United Network for Organ Sharing to manage the mission, with about 22,000 people receiving transplants each year – but there are never enough donors to meet demand. In the UK, 3,740 major organ transplants were carried out in 2010–11, a number that could easily have been twice as high if a sufficient number of donor organs had been available, especially when you consider that these organs came from just over 2,000 donors. In China, more than 13,000 people received kidney or liver transplants in 2004, but the annual number of transplants has fallen since then, mostly due to a lack of donors. In 2009, outrage erupted in the UK when it was revealed that wealthy individuals had been able to jump an 8,000-person-long waiting list to receive vital organ transplants – even though many of the lucky recipients were not even British. EU regulations allow citizens of member states to seek treatment in Britain, and although hospitals are not compelled to admit them, private patients are highly profitable – good for the bottom line. Over the course of two years, fifty individuals from countries as far flung as Greece, Libya, the UAE, China and Israel received livers from British organ donors. The operations took place in private facilities in London, with surgeons able to command a fee of around twenty thousand pounds per operation. Angry headlines prompted a government review into how organ transplants were handled – proving once again that if there is one thing the British cannot abide, it is queue jumping.*

Just as the demand for body parts is plagued by murky dealings, so too is the supply. Consider the case of Sergei Malish, a teenager from Ukraine. At his funeral, Malish's parents discovered deep cuts in the young man's wrists, even though he had died from hanging. Further investigation revealed that Malish's

---

* When pushed to expand on these issues by a reporter from the London *Evening Standard*, kidney specialist Nadey Hakin said that the people who would be most upset by the news were the people giving, not receiving: "Donors will be horrified to hear this is happening."

body had been stripped for parts without their (or his) permission. It was just the latest case in a rash of body-part thefts in the country. Malish's bits and pieces had been bundled off to a German laboratory that refined the raw material into medical implants and relabelled them as being of German origin. The final products were then shipped to the US or South Korea, where they likely would have been used to treat a less-than-life-threatening condition. In this sort of human harvest, the organs that are most valuable are often the skin, which is mashed up, and bones, which are whittled down and used for breast reconstruction, nose jobs and wrinkle treatment. Tiny screws are whittled from bones and used to secure dental and orthopaedic implants, leftover bone is ground into glue. Unscrupulous harvesters replace the bones with PVC piping to conceal the fact that they have been taken. Tendons, heart valves, veins, ribs, eardrums and teeth are all pulled out. Using forged documents, the illicit materials are passed around the world to tissue dealers who often don't know where the material originated from, and don't seem to care too strongly. As well as being a global business, the scale of the trade in body parts is huge. RTI Biologics, one of the companies indicted in the Ukrainian body-parts scandal, reportedly manufactured over half a million implants in 2011, netting $11 million (£7 million) in pre-tax profits.

We should not be surprised that the trade in human tissues is beset by corruption and dubious ethics – that's what happens in any market where demand outstrips supply. We rely on human beings as a resource each and every day, fuelling a world-spanning supply chain. Our demand for cheap products depends on the sweatshops of Asia, the migratory fruit pickers of California, and the gang-masters of Britain; cheap labour supplies us with the stuff of life. And when the stuff of life that a person needs is a functioning liver or kidney, why wouldn't someone be tempted to pay the price for it, or to steal one for a guaranteed profit?

Nearly a hundred years ago, in the face of the punishing labour seen in the fields and factories, a pair of Czech brothers, Karel and Josef Čapek, wondered if there might be a solution that did not involve exploiting our fellow humans. In 1920 they coined the word "robot", from the Czech *robota*, meaning "labour", or more precisely "drudgery". The Čapeks' robots were synthetic serfs, fashioned from a protoplasmic chemical, who would be happy to answer every human need. These weren't metallic machines or silicon-brained androids but physical bodies at our disposal.

What if we were able to grow a multitude of new humans – protoplasmic chemical clones, replicants, tanks, blanks, spares – all quite happy to be stripped for parts, because they are a living body "without a soul"? There would no longer be a shortage of organ donors, or the pesky needs for consent. In other words, could we grow zombies, and keep them alive until we needed them?

LETTUCE HEADS AND ARTICHOKE HEARTS

Sheep rarely make the headlines, but in 1996, a lab at the Roslin Institute celebrated the birth of Dolly, the first mammal to be cloned from an adult somatic cell. The public was at turns enraptured and repulsed with the possibilities that this success-ful cloning wrought. Dead pets might be re-created, and so might dead children; perhaps great artists and thinkers might be duplicated, presuming their abstract talents were genetically encoded. Bill Clinton was moved to point out that the cloning success raised "serious ethical questions, particularly with respect to the possible use of this technology to clone human embryos". Later that year, he imposed a pre-emptive five-year ban on cloning humans.

Soon after Dolly's birth announcement, the BBC's flagship science programme *Horizon* set out to investigate the implications of the new technology. The producers got in touch with Jonathan Slack, a professor of developmental biology at the University of Bath. Slack hadn't been involved in the sheep project, but he was a prominent scientist working in genetics, and he was a perfect candidate to comment. During a preliminary conversation, Slack mentioned in passing that the procedure that had created Dolly opened up the possibility of growing specialized organisms. "Maybe if our knowledge of how to suppress body parts were combined with the cloning technology", he mused, "we could grow human organs for transplantation". It was a train of thought he would live to regret voicing.

The BBC crew decamped to the lab for a day of recording, conducting a lengthy conversation with Slack and taping his assistants while they were engaged in visually attractive but fairly meaningless bits of "science", such as pipetting water from one test tube to another and injecting blue dye into cells (it's not known if it was Niagara sky blue 6B). One part of the work carried out in Slack's lab looked at the effects of suppressing various genes in frogs. By tweaking the genetic structure of developing eggs, the resulting tadpoles could be made to grow without heads or without tails – or as just a head or a tail suspended in a drop of gelatinous frogspawn. In this way, the scientists could probe the role of specific genes in the embryonic development of a range of vertebrates, including humans. Recalling the visit in his book *Egg and Ego*, Slack notes:

> The science fiction, or informed speculation, went like this: suppose someone needed an organ transplant. You could culture some of his cells. They could be any type of cells, as the genes in all cells are the same; so white blood cells would do. Genes would then be introduced into the cells while they were growing in culture that would have the effect of suppressing the development of most parts of the embryo, except for the part of the body that you want... A genetically modified cell would then be fused with an enucleated human egg, and the resulting reconstituted embryo grown up, preferably in vitro, as an "organ culture". Particularly if the culture could be nourished at a higher rate than a normal embryo, it would grow to transplantable size within months. The patient would then have an organ graft that was a perfect genetic match and required no immunosuppression.

When a *Sunday Times* journalist who had previewed the film called to speak to Slack, he repeated his comments. The newspaper decided to run the story on their front page, under the banner: HEADLESS FROG OPENS WAY FOR HUMAN ORGAN FACTORY. Slack was forced to spend the following weeks fending off news teams from around the world, repeatedly stating that he had not made headless frogs (only tadpoles!), and that he was not preparing to grow decapitated human vegetables from which to harvest organs.

Slack was far from the first person to foresee the development of lab-grown beings that might supply us with organs for much-needed transplants. Russian surgeon Vladimir Demikhov had shocked the world when he debuted a two-headed dog in 1954. The animal was created by grafting the head, neck and front legs of a puppy onto the body of an adult mastiff. The puppy's body parts shared a blood supply with the adult, but the dogs had little else in common – the two heads ate (one in vain), slept and acted independently of each other. The puppy even enjoyed aggravating its host by nibbling on the mastiff's ears. Demikhov created several of these double dogs, but infection and tissue rejection meant that the bicephalic beasts never lasted long. Demikhov persevered, but the most successful survived for a brief twenty-nine days. Most observers branded the work as a stunt, but for Demikhov it was an essential step towards his ultimate goal: a human heart-lung transplant.

Along the way, Demikhov envisioned surgeries during which he would attach extra limbs and organs to brain-dead humans. He called these graft recipients "human vegetables", but it isn't clear if by this he was referring to their unconscious state or to some potential to sprout new appendages. In any case, the "vegetables" would supply and nurture the surplus organs until such a time as they were needed, whereupon doctors could pick them like ripe fruit for use in transplants. In the future, Demikhov hoped, human organs would be farmed in every hospital and research laboratory.

Due to problems with tissue rejection, the plan was completely unfeasible, but the idea of using the perfect vessel for sustaining spare organs didn't die with Demikhov. Indeed, in *Egg and Ego*, Slack suggests somewhat mischievously that women might be employed to grow new organs for those in need:

> The least plausible part of the scenario is the growth of the organ culture in vitro. In vitro culture of mammalian embryos is painfully difficult, and we are a long way from being able to support any type of mammalian embryo through gestation outside the mother. On the other hand, fertilised eggs are being implanted in the wombs of women every day, so my guess is that when the time comes, female volunteers will be needed to incubate the organ cultures. This could be an act of love, to grow an organ for a relative who needs one. But it could also be an act of commerce, and the unseemly rows arising from existing maternal surrogacy cases make one doubtful about how wise it would be to go down that road.

Neither Demikhov's human vegetables nor Slack's organ incubators are likely to see the light of day, mostly because, since Slack made his suggestion in 1998, scientists have become quite adept at growing whole organs in the lab. At the Wake Forest Institute for Regenerative Medicine in Ohio, researchers hacked an ordinary inkjet printer, filling the ink cartridges with epidermal cells cultured from a patient. Using this, they were able to "print" 3D skin grafts directly onto the wounds of burn victims, reducing healing time and eliminating the need to remove patches of skin from elsewhere on the body to be used as a graft. A similar 3D-printing technique has also been used by scientists at Newcastle University to create a human liver out of stem cells. Bladders, kidneys, lungs, bones, ears, ligaments, corneas and ovaries have all been produced in Petri dishes. And in 2008, British surgeons carried out the first transplant of a mature, lab-developed organ, replacing the damaged windpipe of thirty-year-old Claudia Castillo with one created by growing Castillo's cells on a biological scaffold of a donor trachea.

Our skill in growing individual organs has progressed rapidly, and much sooner than many scientists anticipated just a generation ago. Top-down organ harvesting – growing and nourishing an entire adult clone so that you might later take out the parts you need – would be time-consuming, energy intensive and wasteful by comparison. Much more efficient to simply build the bit you need, when you want it.

There's only one exception to that cost-benefit analysis – when you want a whole human.

HAPPY ENDINGS

On 5 April 2009, Nikolas Evans was walking home from a bar in Austin, Texas, with a friend. Before they reached the bus stop they were attacked, and in the scuffle Evans was knocked to the ground, hitting his head hard on the asphalt. He was left lying unconscious in the road. Ten days later he died of a subdural haematoma, a type of swelling inside the skull that crushes the brain. A year later, however, the deceased Evans was about to start a family.

Evans had died without leaving a partner, and neither had he made any donations to a sperm bank; his very late entry into parenthood came courtesy of his distraught mother, Missy, who arranged for Evans's sperm to be collected after the twenty-one-year-old's death. While the young man lay in a coma, his mother floated her plan by other family members, all of whom offered their support. She then secured permission from a probate judge to take ownership of her son's sperm. His body was chilled and the samples were extracted by a sympathetic doctor who provided her services free of charge. Five of Evans's organs were also removed, passed on to needy individuals for transplant.

The operation was, to say the least, unusual, and received attention in the local press. After the news broke, Evans's mother said she was contacted by "hundreds" of women offering to donate eggs or be surrogate mothers to her son's posthumously conceived grandchild. She was also subjected to intense criticism: attacked for being unmarried and un-Christian, told that she was selfishly trying to produce a child to replace the one she had lost. She remained resolute: "He helped five people live that day. Why can't I have a gift? Why do I lose everything?"

The first case of post-mortem sperm retrieval had been carried out twenty years earlier, on a thirty-year-old killed in a road traffic accident. The man's family had requested that he be maintained on life support so that his sperm could be collected. The procedure is relatively simple: urologists can extract sperm surgically, by slicing directly into the testes, or by triggering ejaculation with an electric rectal probe, putting a macabre twist on the ages-old massage parlour promise of a "happy ending". The first child born to a post-mortem donor did not arrive until much later, in 1999, when Bruce Vernoff's daughter was born two years after his sudden death from an allergic reaction.

It's not just dead men who've been busy making babies. A decade after the birth of Louise Joy Brown, the first "test-tube baby", a research team at the Harbor-UCLA Medical Center in California successfully transferred a fertilized egg from one woman to another, resulting in the birth of a bouncing baby boy nine months later. Although the technology has improved considerably since then (of forty-six transfers attempted by the 1984 team, only two were successful), sourcing eggs is still anything but straightforward. The donor, usually a young woman in her twenties, undergoes a month-long course of injected hormone treatments, then is anaesthetized so that a bounty of eggs can be removed with a syringe-like needle. The process if much more time consuming and invasive than masturbating into a sample cup – the experience of most sperm donors. The onerous demands on egg donors account for both the high price tag (around six thousand US dollars per round) and the relative scarcity of donated eggs.

Just as clinics looked to the deceased when faced with a shortage of blood eighty years ago, so they have looked to the deceased as a potential source of much-needed eggs. One survey of seven hundred Utah residents plumbed people's attitudes towards the use of genetic material taken from a brain-dead organ donor. More than 70 percent supported the use of ovary tissue for scientific research, but when it came to reproduction, opinion swung the other way. Most said it was less acceptable to fertilize eggs extracted from the dead to create pre-embryos, let alone to implant those embryos in women desperate to become pregnant. In particular, the respondents reported that they would not be comfortable directing the process as a surviving legal guardian of the deceased woman. The implication: if a pregnancy occurs, it should be with the express instructions of the woman. But what young woman plans to die before she has a chance to start a family, or thinks to write a will giving her eggs to infertile couples? Even if she signed up to be an organ donor, would she intend for that to include her eggs?

If the eggs are kept "in the family", the issues are no less simple. In 2010, Dr Anna Smajdor of the University of East Anglia reported on the first case of an attempt to retrieve eggs at the point of death – a "peri-mortem" retrieval. Seven hours into a long-haul flight, a woman suffered a pulmonary embolism, a blood clot that triggered a heart attack and left her brain damaged. After an emergency landing in Boston, she was rushed to a hospital and put on a ventilator. She was not quite brain dead, but her condition was gradually declining, with little hope of recovery. Her husband and family were consulted, and they agreed to scale down the woman's treatment. Subsequently, they had a change of heart, asking for the patient to be kept alive long enough to organize the retrieval of some eggs, which could be fertilized by her partner to conceive a child. The embryo could then be implanted into a surrogate mother who would carry the offspring to full term. The request threw the doctors into a quandary. Theoretically, the procedure was possible, even easy. But no one had ever tried it before, making it uncharted territory medically, legally and ethically.

There was no sign of consent or advance planning from the patient – how could they know what she wanted? Would the egg-harvesting procedure hasten the patient's death? More importantly, was there even a convincing medical justification for the use of a surrogate? As Smajdor writes:

> The need for egg harvesting could be avoided altogether, just as the need for surrogacy could. The

husband could attempt to impregnate his comatose
wife in the "natural" way. We might blench at the
idea of this, but if so, we must ask why. The answer
is likely to be that, in the absence of consent, the act
of sexual intercourse is usually a serious crime. But
precisely the same can be said of surgical intervention. If we can countenance one without consent,
then why not the other? Certainly, if a pregnancy
could be achieved more cheaply and more safely
through intercourse than through a complex series
of surgical interventions, it is hard to see why the
latter should be preferred.

Doctors discussed the request with the assisted reproductive
team, who advised them that the patient would require at least
two weeks of hormone therapy to stimulate her ovaries before
eggs could be collected. In addition, the procedure itself would
require lying the patient flat on her back, which would likely
result in brain herniation and death. Ultimately, the doctors
declined the family's wishes. The patient was taken off life support and passed away soon thereafter.

There are numerous documented cases of critically ill women
who have been kept alive long enough to complete a pregnancy,
women who are on the cusp of death ushering a new life into
the world. When Chastity Cooper, a twenty-four-year-old
woman from Kentucky, crashed her car in bad weather on her
way home one night, no one knew she was pregnant. Routine
tests carried out at Cincinnati University Hospital revealed that
she had conceived just two weeks before her accident. Sustained
on life support, she was able to carry the pregnancy to full

term, giving birth naturally thirty-six weeks later to one Alexis Michelle Cooper, a healthy girl weighing in at seven pounds, seven ounces. Cooper's attending physician, Dr Michael Hnat, remarked on the unusual nature of the pregnancy: "This is one of the only cases ever in the United States where the woman was in a coma throughout the entire gestation."

During her pregnancy, Cooper's condition had improved gradually, to the point where she was able to open her eyes; her gaze would follow visitors around the hospital room. After the delivery, her husband (and new father) Steve told the gathered reporters: "It was like magic. When the baby came out, Chastity just smiled." The family hoped a second miracle would appear – that the young woman would wake from her coma. It was a long shot; Cooper's husband, despite losing his job and his house during his wife's illness, was prosaic: "I try not to think about how hard it is," he said. "I'm a firm believer that you have to play the cards you're dealt."

There have been at least eleven children born to irreparably brain-damaged women in the US since 1979, a number that will only grow. Smajdor warns: "Such requests are likely to emerge more frequently as technology offers new avenues for the pursuit of parenthood, and a robust, simple legal approach is vital."

That we are able to draw sperm from dead men, or keep brain-dead women alive to bring a pregnancy to full term, is amazing. But, ultimately, so is every new medical triumph, and we've had quite a number of those over the past fifty years. As a society, we've become jaded towards technological progress, at times even flat-out disappointed. The future hasn't lived up to its promises – I still haven't received my jetpack – and so we invest our passion in *how* technologies are used, not in what technologies we *have*. If we know that a person would have wanted to have children, and the procedure to extract his or her gametes is possible and carries no detrimental effect for others, we should do it. If we can keep a woman alive artificially in order to save her unborn child, we should do that, too. Those debates seem closed. The debate over whether the consent given by whole-body organ donors includes their gametes is the next frontier. If we can countenance a woman becoming pregnant peri-mortem with her own child, and carrying it to term, where do we stand on brain-dead organ donors acting as surrogates for other people's children? Given the risks of pregnancy to the mother, surely this is preferable to using living surrogates, especially in light of blossoming "baby-farms" in India where young women are paid – some would say exploited – by wealthy foreigners who want someone to gestate their offspring for them? Will we ever see organ donation move beyond the gifting of individual organs and tissues and into the services of the body itself? Will the organ donors of the future be quiet, dependable producers of blood, plasma, gametes and hormones – even incubators of unborn children – who are little more than some cells hung together to suit our needs? Isn't that what we call meat?

DINING WITH THE DEAD

When William Seabrook sat down at a dinner table in Paris and tucked into his most infamous meal, he joined the ranks of a very small group of humans who have eaten human flesh not out of desperation, but simply because they wanted to. Despite being held up as an ultimate taboo, the practice of cannibalism was once geographically and culturally widespread, occurring from the far reaches of the South Pacific to the very heart of Britain and many places in-between. Those who write the history books are often keen to downplay any suggestion that their own people would be capable of "such a thing". For instance, cut marks found on bones in human burial sites are written off as "ritual funerary defleshing" without a mention of where that meat ended up. Something had to consume it, and you can be certain that it wasn't always bacteria and larvae.

There are a few cultures for which it is known that cannibalism has been a regular ritual (not simply an "occasional" medical intervention, as among those blood-slurping epileptic Europeans). The Fore people of Papua New Guinea ate their dead relatives as part of funeral rites; cannibalism was said to keep the vital essence of the deceased within the community. Unfortunately, something else was retained in the bargain: a deadly neurological disorder known as kuru, related to Creutzfeldt-Jakob or "mad cow" disease, in which proteins become misfolded and infect surrounding tissue, causing tiny holes like in a sponge (the diseases are known as spongiform). People struck by kuru grow steadily weaker, acquiring tremors in their hands and feet. They walk with jerky, uncertain steps, their voice becomes slurred and their moods grow unstable. They laugh spontaneously without understanding why. Eventually they lose the ability to stand up, speak or swallow food, with death shortly following. During the cannibalistic feasts of the Fore people, women and children typically ate the dead person's brain, where the infectious prion particles were most concentrated, and thus they were eight times more likely to be felled by the disease than if they ate other parts of the body. When cannibalism became criminalized under the law, the disease vanished – though there is some debate over whether it was already in decline, with survivors more likely to carry a genetic variant that is resistant to prions.

The kuru outbreak is a cautionary tale against thinking of our fellow humans as a safe and handy food source, but for all of the billions of people on the planet, only a satirist like Jonathan Swift would argue that the sheer numbers make it very tempting. The consumption of humans makes little ecological sense – as we've already seen, we're slow-growing, energy-intensive and too damn scrawny. As meat, humans just don't compare to pigs and cows, so the prospects of wolfing down chips of Soylent Green are pretty slim.

Still, that doesn't mean we're completely off the menu. In 2011, an ice-cream parlour in London's Covent Garden began selling Baby Gaga, a Madagascan-vanilla- and lemon-zest-infused ice cream made with human breast milk. Vending at a cool £14 (US $22.50) per scoop, the unusual treat managed to stay on sale for less than a week before stocks were seized by health officers. (Any and all snacks, the authorities reminded London's foodies, "made from another person's bodily fluids can lead to viruses being passed on and, in this case, potentially hepatitis".) Every man or woman to his taste.

What we are willing to ingest says a lot about what it means to be human. Perhaps we'll never feel comfortable raising a six-pack of humanlike zombies purely to consume them. But if we did, the other part of Karel Čapek's robotic future might come to be: the artificial serfs become unhappy and rebel, exterminating humanity.

Epilogue
HERE AND NOW

I'M NOT JOKING ABOUT THAT ZOMBIE UPRISING.

When I started this book, I thought that I might find a case that really pushed the boundaries of death and miraculous recovery, or a surprising tale of somebody, brain laced with tungsten filament and drugs, taken under the control of another person. I thought that the ingenuity of the human race would provide the tools to make a zombie, or at least something approaching it. The reality was something far more disturbing. Instead of uncovering the scientific tools that might overcome death, I found that death itself is a fuzzy principle – the only certainty being that, so long as our brains and bodies can replace cells faster than they die, we're considered to be alive. If you stick with the science, death stops being a line you cross and becomes instead a matter of margins: how much damage can your brain and body tolerate before the intricate clockwork of your body clatters and hiccups to a standstill?

Life is wearing us away all the time – that's just how biology works. Like the 'ship of Theseus', we're replaced cell by cell until several years down the line not a single original part remains. Even our brains are a restless sea of interlaced neurons, constantly shifting in blind eddies. Yet somehow some piece of it, our *ti-bon anj*, manages to step daintily into the flesh of those new cells, new arrangements. Or as Hammer Horror might have put it, an eternal spirit flickering inside a lantern of slowly decaying flesh.

But before you get too philosophical about humanity's permanent state of death and decay, remember that the mind is corruptible. And indeed, all of our mental activity dissolves into flickers of light too – patterns cast by the arrangement of neurons and dendrites and axons as they set spark to the mind. So while genuine mind control has yielded only scant progress thus far (in humans), it's not because the mind is particularly resistant to outside control. Your personality, your id, the very essence of you, is nothing more than the sophisticated interplay of uncountable cells and chemicals – and some of those cells and chemicals may not be your own. Could you say where your identity ends and the influence of a parasite begins? Forget controlling other people's minds – we can't even lay claim to our own. Maybe one day we'll appropriate bugs such as *T. gondii* as a tool for our well-being, like we did with the bacteria in our gut, slurping down probiotic yoghurt to complement our psychomicrobiome.

The science of zombies reveals that life is not discrete. Each of us is a confluence of many billions of cells, none of which are entirely under your control or entirely living. When you die, these material fibres of what we call the soul will be scattered to the four corners of the Earth, to live on in perpetuity. You are an undead zombie, and you always have been.

SELECTED BIBLIOGRAPHY

Introduction: Recipe for a Zombie

n.a., (2003) 'Arizona man keeps wife's remains in freezer for years', Associated Press 13 September

Pela, Robrt L. (2003) 'Bitter end', *Phoenix New Times* 2 October, http://www.phoenixnewtimes.com/2003-10-02/culture/bitter-end/

1. Dead Men Working the Fields

Berlinski, Mischa (2009) 'Into the zombie underworld', *Men's Journal* 17 September

Birmingham, A.T. (1999) 'Waterton and Wourali'a', *British Journal of Pharmacology* 126 (8): 1685–90

Davis, Wade (1983) 'The Ethnobiology of the Haitian zombie', *Journal of Ethnopharmacology* 95 (November): 85–104

Davis, Wade (1988) *The Serpent and the Rainbow*, London and Glasgow: William Collins Sons & Co. Ltd

Davis, Wade (1988) 'Zombification', *Science* 240 (24 June): 1715

Hearn, Lafcadio (1903) *Two Years in the French West Indies*, Harper & Brothers Publishers, http://www.gutenberg.org/ebooks/6381

Hoffman, Bill (2005) 'Blood swapping reanimates dead dogs', *New York Post* 28 June

Lee, Cheng Chi (2008) 'Is human hibernation possible?', *Annual Review of Medicine* 59: 177–86

Littlewood, Roland (2009) 'Functionalists and zombis: Sorcery as spandrel and social rescue', *Anthropology and Medicine* 16 (3) (December): 241–52

Littlewood, Roland, and Chavannes Douyon (1997) 'Clinical findings in three cases of zombification', *Lancet* 350 (11 October): 1094–6

NewsCore (2010) 'Baby cooled for four days to fix heart condition', Fox News 17 June, http://www.foxnews.com/health/2010/06/17/baby-cooled-days-fix-heart-condition/

Page, Lewis (2010) 'Suspended-animation cold sleep achieved in lab', *Register* 11 June, http://www.theregister.co.uk/2010/06/11/suspended_animation_in_lab/

Safar, Peter (2000) 'On the future of reanimatology', *Academic Emergency Medicine* 7 (1): 75–89

Safar, P., S.A. Tisherman, et al. (2000) 'Suspended animation for delayed resuscitation from prolonged cardiac arrest that is unresuscitable by standard cardiopulmonary-cerebral resuscitation', *Critical Care Medicine* 28 (November supplement): N214–8

Seabrook, William B. (1929) *The Magic Island*, New York: Harcourt Brace and Company

Stark, Peter (1997) 'The cold hard facts of freezing to death', *Outside* (January)

Waterton, Charles (1838) *Essays on Natural History, Chiefly Ornithology. With an Autobiography of the Author*, London: Longman, Brown, Green and Longmans

Wood, Clair G. (1987) 'Zombies', *ChemMatters*, 4

Wu, X. et al. (2008) 'Emergency preservation and resuscitation with profound hypothermia, oxygen, and glucose allows reliable neurological recovery after 3 h of cardiac arrest from rapid exsanguination in dogs', *Journal of Cerebral Blood Flow and Metabolism* 28 (2): 302–11

2. Time for a Revival

Appleyard, Sam (2008) 'The living dead', *Sunday Times* 14 December

Aynsley, E.E. and W.A. Campbell (1962) 'Johann Konrad Dippel 1673–1734', *Medical History* 6 (3): 281–6

Banner, Stuart (2003) *The Death Penalty: An American History*, Cambridge: Harvard University Press

Barrett, Sam (2008) 'The first few minutes after death', PopSci 31 October, http://www.popsci.com/sam-barrett/article/2008-10/first-few-minutes-after-death

Bartoll, Jens (2008) 'The early use of Prussian blue in paintings', Ninth International Conference on NDT of Art, Jerusalem, Israel, 25–30 May

Cornish, Robert E. and H.J. Henriques (1933) 'Report of investigation of resuscitation', unpublished, 8 October

Cornwall, J.W. (1935) 'Jiu-jitsu methods of resuscitation', Correspondence, *British Medical Journal* 2 (3893): 17 August

Duffy, Clinton T. (1950) *The San Quentin Story As Told to Dean Jennings*, Garden City, NY: Doubleday

Ford, J.E. (1935) 'Can science raise the dead?', *Popular Science Monthly* February

'George Foster', *Proceedings of the Old Bailey, London's Central Criminal Court, 1674 to 1913*, Old Bailey Online (accessed 29 June 2010)

Krementsov, Nikolai (2009) 'Off with your heads: Isolated organs in early Soviet science and fiction', *Studies in History and Philosophy of Biological and Biomedical Sciences* 40 (2): 87–100

n.a. (1903) 'Revival of isolated heart after death', *Journal of the American Medical Association* (21 March)

n.a. (1929) 'Artificial heart keeps dog's head alive for hours', *The Tech* 20 February n.a. (1934) 'Lazarus, dead & alive', *Time* 26 March

n.a. (1935) 'Scientist to make bold attempt to revive human dead', *Modern Mechanix* February, reprinted at http://blog.modernmechanix.com/ scientist-to-make-bold-attempt-to-revive-human-dead/

O'Donnell, C.P.F., A.T. Gibson and P.G. Davis (2006) 'Pinching, electrocution, ravens' beaks, and positive pressure ventilation: A brief history of neonatal resuscitation', *Archives of Disease in Childhood: Fetal and Neonatal* 91 (5): F369–73

Paris, John Ayrton and John Samuel Martin Fonblanque (1823) 'The Application of the physiological facts established in the preceding chapters, to the general treatment of asphyxia', *Medical Jurisprudence*, London: W. Phillips

Parnia, Sam (2007) 'Do reports of consciousness during cardiac arrest hold the key to discovering the nature of consciousness?', *Medical Hypotheses* 69 (4): 933–7

Stafford, Jane (1934) 'Can the dead be given life?', *Science News-letter* 1 December

Wilkes, John (1810) 'John Conrad Dippel', *Encyclopaedia Londinensis*, http://archive.org/details/encyclopaedialon15wilk

3. Mickey Finn and Other Thugs

Albarelli, Hank P. (2009) *A Terrible Mistake: The Murder of Frank Olson, and the CIA's Secret Cold War Experiments*, Walterville, OR: Independent Publishers Group

Blakeslee, Sandra (2005) 'This is your brain under hypnosis', *New York Times*, 22 November

Gabbai et al. (1951) 'Ergot poisoning at Pont St. Esprit', *British Medical Journal* 2 (15 September): 650–1

Hooper, Judith and Dick Teresi (1991) *The Three-Pound Universe: Revolutionary Discoveries about the Brain – from the Chemistry of the Mind to the New Frontiers of the Soul*, London: J.P. Tarcher

Jay, Mike (1999) *Artificial Paradises*, London: Penguin Books

n.a. (2005) 'Restaurant shift turns into nightmare', ABC News *Primetime* 10 November, http://abcnews.go.com/Primetime/story?id=1297922&page=1

Proenza, Anne (1994) 'Losing their minds in Bogota', *World Press Review* 41 (10): 20–1

Thomson, Mike (2010) 'Pont-Saint-Esprit poisoning: Did the CIA spread LSD?', BBC News, http://www.bbc.co.uk/news/world-10996838

Raz, Amir, J. Fan et al. (2005) 'Hypnotic suggestion reduces conflict in the human brain', *Proceedings of the National Academy of Sciences USA* 102 (28): 9978–83

Zetter, Kim (2010) 'This day in tech: April 13, 1953: CIA OKs MK-ULTRA mind-control tests', *Wired*, 13 April, http://www.wired.com/thisdayintech/2010/04/0413mk-ultra-authorized

4. Remote / Control

Blackwell, Barry (2011) 'José Delgado' [obituary], *American College of Neuropsychopharmacology*. Available at www.acnp.org/asset.axd?id=f9da6400-ea5f-4b24-990e-6c259d48eca4 (Accessed 26 February 2013)

Costandi, Mo (2006) 'The incredible case of Phineas Gage', *Neurophilosophy* 4 December, http://neurophilosophy.wordpress.com/2006/12/04/the-incredible-case-of-phineas-gage/

El-Hai, Jack (2007) *The Lobotomist: A Maverick Medical Genius and His Tragic Quest to Rid the World of Mental Illness*, Hoboken, NJ: Wiley (2007)

Horgan, John (2005) 'The forgotten era of brain chips', *Scientific American* October: 66–73

n.a. (1951) 'Grey matter', *Time* 28 May, http://www.time.com/time/magazine/article/0,9171,890110,00.html

n.a. (1952) 'Mass lobotomies', *Time* 15 September, http://www.time.com/time/magazine/article/0,9171,816987,00.html

Nuzzo, Regina (2008) 'Call him doctor "Orgasmatron"', *Los Angeles Times* 11 February, http://www.latimes.com/features/health/la-he-orside11feb11,0,79450.story

Ravo, Nick (1999) 'Robert G. Heath' [obituary], *New York Times* 25 September, http://www.nytimes.com/1999/09/25/us/robert-g-heath-84-researcher-into-the-causes-of-schizophrenia.html

Young, Robert M. (1970) *Mind, Brain and Adaptation in the Nineteenth Century: Cerebral Localization and Its Biological Context from Gall to Ferrier*, Oxford: Clarendon Press

5. The Ghoulish Nanny

Adamo, S., C. Linn and N. Beckage (1997) 'Correlation between changes in host behaviour and octopamine levels in the tobacco hornworm *Manduca sexta* parasitized by the gregarious braconid parasitoid wasp *Cotesia congregata*', *Journal of Experimental Biology* 200: 117–27

Amos, Jonathan (2000) 'Parasite's web of death', BBC News, 19 July, http://news.bbc.co.uk/1/hi/sci/tech/841401.stm

Costandi, Mo (2006) 'Brainwashed by a parasite', *Neurophilosophy* 20 November, http://neurophilosophy.wordpress.com/2006/11/20/brainwashed-by-a-parasite/

Fisher, Roderick C. (1961) 'A study in insect multiparasitism', *Journal of Experimental Biology* 38: 267–75

Grosman, Amir H. et al. (2008) 'Parasitoid increases survival of Its pupae by inducing hosts to fight predators', *PLoS ONE* 4 June, http://www.plosone.org/article/info:doi/10.1371/journal.pone.0002276

Haspel, Gal, Lior Ann Rosenberg and Frederic Libersat (2003) 'Direct injection of venom by a predatory wasp into a cockroach brain', *Journal of Neurobiology* 56 (3): 287–92

Holmes, Bob (1993) 'Evolution's neglected superstars', *New Scientist* 6 November, http://www.newscientist.com/article/mg14018983.500–

evolutions-neglected-superstars-there-is-nothing-glamorous-about-fleas-flukes-or-intestinal-worms-so-why-are-they-suddenly-attracting-so-much-attention.html

Jog, Maithili and Milind Watve (2005) 'Role of parasites and commensals in shaping host behaviour', *Current Science* 89: 7: 1184–91

Sapolsky, Robert (2003) 'Bugs in the brain', *Scientific American* March: 94–7

Zimmer, Carl (2002) *Parasite Rex: Inside the Bizarre World of Nature's Most Dangerous Creatures*, New York: Simon & Schuster

6. Army of Bloodsuckers

Botto-Mahan, Carezza, Pedro E. Cattan and Rodrigo Medel (2006) 'Chagas disease parasite induces behavioural changes in the kissing bug *Mepraia spinolai*', *Acta Tropica* 98 (3): 219–23

Callahan, Gerald N. (2002) 'Infectious madness: Disease with a past and a purpose: Mental illness may not be just craziness, but have a parasitic, fungal, or viral etiology', *Emergency Medicine News* 24 (11): 52–4

'Edgar Allan Poe mystery', press release, University of Maryland Medical Center, Baltimore, MD (24 September 1996), http://www.umm.edu/news/releases/news-releases-17.htm

Flegr, Jaroslav et al. (2002) 'Increased risk of traffic accidents in subjects with latent toxoplasmosis; a retrospective case-control study', *BMC Infectious Diseases* 2 (2 July): 11

Lacroix, Renaud et al. (2005) 'Malaria infection increases attractiveness of humans to mosquitoes', *PLoS Biology* 9 August, http://www.plosbiology.org/article/info:doi/10.1371/journal.pbio.0030298

Lefèvre, T. et al. (2007) '*Trypanosoma brucei brucei* induces alteration in the head proteome of the tsetse fly vector *Glossina palpalis gambiensis*', *Insect Molecular Biology* 16 (6): 651–60

n.a. (2010) 'A game of cat and mouse', *The Economist* 3 June, http://http://www.economist.com/node/16271339

Schultz, Nora (2007) 'Zombie cockroaches revived by brain shot', *New Scientist* 30 November, http://www.newscientist.com/article/dn12983-zombie-cockroaches-revived-by-brain-shot.html

Thomas, F. et al. (2002) 'Do hairworms *(Nematomorpha)* manipulate the water seeking behaviour of their terrestrial hosts?', *Journal of Evolutionary Biology* 15 (30 April): 356–61

Webster, J.P. et al. (2006) 'Parasites as causative agents of human affective disorders? The impact of anti-psychotic, mood-stabilizer and anti-parasite medication on *Toxoplasma gondii*'s ability to alter host behaviour', *Proceedings of the Royal Society, Biological Sciences* 273 (1589): 1023–30

Yereli, Kor, I. Cüneyt Balcioglu and Ahmet Ozbilgin (2006) 'Is *Toxoplasma gondii* a potential risk for traffic accidents in Turkey?', *Forensic Science International* 163 (10 November): 34–7

7. The Human Harvest

Bell, Vaughan (2010) 'Gladiator's blood as a cure for epilepsy', MindHacks 1 February, http://mindhacks.com/2010/02/01/gladiators-blood-as-a-cure-for-epilepsy/

Currell, Susan and Christina Cogdell, eds (2006) *Popular Eugenics: National Efficiency and American Mass Culture in the 1930s*, Athens: Ohio University Press

Dawson, Warren R. (1927) 'Mummy as a drug', *Proceedings of the Royal Society of Medicine* 21 (1): 34–9

Friedman, David M. (2008) *The Immortalists: Charles Lindbergh, Dr Alexis Carrel, and Their Daring Quest to Live Forever*, New York: Ecco

James, Susan Donaldson (2010) 'Sperm retrieval: Mother creates life after death', ABC News, 23 February, http://abcnews.go.com/Health/Wellness/mother-murdered-son-hopes-create-grandchild-post-mortem/story?id=9913939#.TvFUF1auFpk

Kahn, Jennifer (2003) 'Stripped for parts', *Wired* 11 March, http://www.wired.com/wired/archive/11.03/parts.html

Moore, Charles L., John C. Pruitt and Jesse H. Meredith (1962) 'Present status of cadaver blood as a transfusion medium: A complete bibliography on studies of postmortem blood', *Archives of Surgery* 85 (3): 364–70

n.a. (1961) 'Blood from the dead', *Time* 26 May, http://www.time.com/time/magazine/article/0,9171,872489,00.html

n.a. (1973) '"Reprocessed" bodies plan', *Guardian* 10 August, http://www.guardian.co.uk/theguardian/2010/aug/10/archive-reprocessed-bodies-plan-1973

Park, Alice (2007) 'The science of growing body parts', *Time* 1 November, http://www.time.com/time/health/article/0,8599,1679115,00.html

Reggiani, Andrés Horacio (2006) *God's Eugenicist: Alexis Carrel and the Sociobiology of Decline*, New York: Berghahn Books

Shears, Richard, and Rob Cooper (2010) 'Thousands of pills filled with powdered human baby flesh discovered by customs officials in South Korea', *Daily Mail* 7 May, http://www.dailymail.co.uk/news/article-2140702/South-Korea-customs-officials-thousands-pills-filled-powdered-human-baby-flesh.html

Slack, J.M.W. (1998) *Egg and Ego: An Almost True Story of Life in the Biology Lab*, New York: Springer

Taylor, Timothy (2001) 'The edible dead', *British Archaeology* 59 (June), http://www.archaeologyuk.org/ba/ba59/feat1.shtml

Willson, Kate et al. (2012) 'Human corpses harvested in multimillion-dollar trade', *Sydney Morning Herald*, 17 July, http://www.smh.com.au/opinion/political-news/human-corpses-harvested-in-multimilliondollar-trade-20120717-2278v.html

ACKNOWLEDGEMENTS

None of this would have been possible without the combined efforts of too many people to name, here are just a few of them. My agent Peter Tallack of the Science Factory, who took a chance on the book, and fellow zombiephile Marsha Filion at Oneworld who commissioned it. I'm blessed to have Robin Dennis as my editor, who performed alchemy on every draft. A debt of gratitude is also owed to the army of friends, colleagues and strangers who offered time and insight and who tracked down obscure references for me: Aarathi Prasad, Vaughan Bell, Emilia Brock, Jane Bramhill, Andrew Holding, Amanda Hargreaves, Jonathan Slack, Kevin Fong, Stephan Hensel, Marc Mulhern, Andy Reeves, Jamie Gallagher, Dr Becca, Gimpy, Katie Firth, Tom B. Cannon, James Streetley, Sven Rudloff, Hectocotyli, Debayan Sinharoy, Jonathan Parienté, Melissa L. Braaten, Rebecca Dyson, Tommy Leung, Roland Littlewood, Aisling Spain, Peter Cummings, Dr Aust, Stephen J. Henstridge, Beverley Gibbs and many more; and especially Jesus Rogel for his translation work and the Berkeley student who dug out Robert Cornish's original manuscript for me.

Finally, a very special thanks to all my friends, family and my #1 nerd for their limitless patience with me as I wrote this book.

INDEX

ABOUT THE AUTHOR

Frank Swain is the founder of *SciencePunk*, the popular SEED ScienceBlogs site devoted to the weird and wonderful fringes of science. As a science writer, he is preoccupied with how our innovations shape our future and ourselves. His work has appeared in *New Scientist*, *Arc*, *Slate*, *Stylist*, *Wired*, the *Guardian*, *Eureka* and more, and on BBC Radio 4 and Bravo. He lives in London, and this is his first book.